Titanium Dioxide - Uses, Applications, and Advances

Edited by Carlos Montalvo Romero, Claudia Aguilar and Edgar Moctezuma

Published in London, United Kingdom

Titanium Dioxide - Uses, Applications, and Advances
http://dx.doi.org/10.5772/intechopen.1003435
Edited by Carlos Montalvo Romero, Claudia Aguilar and Edgar Moctezuma

Contributors
Abimbola Jacob Olasoji, Alaa H. Taha, Alfredo Cristobal Salas, Ali A. Hassan, Bruna Moura, Bárbara R. Gomes, Carolina Solis Maldonado, Edwin Mapasha, Eric Maluta, Ife Elegbeleye, Ioana Stanciu, José Luis Xochihua Juan, Leskey Mduduzi Cele, Lorena Coelho, Mariana Ornelas, Martina Kocijan, Matejka Podlogar, Miklós Fried, Nayeli Ortiz Silos, Noor Taha Ismaeel, Peter Petrik, Raúl Alejandro Luna Sánchez, Regina Maphanga, Sang Hyuk Im, Tatiana L. Izaguirre Gallegos, Zoltán Lábadi

First published in London, United Kingdom, 2025 by IntechOpen
IntechOpen is the global imprint of INTECHOPEN LIMITED, registered in England and Wales, registration number: 11086078, 167-169 Great Portland Street, London, W1W 5PF, United Kingdom

For EU product safety concerns: IN TECH d.o.o., Prolaz Marije Krucifikse Kozulić 3, 51000 Rijeka, Croatia, info@intechopen.com or visit our website at intechopen.com.

British Library Cataloguing-in-Publication Data
A catalogue record for this book is available from the British Library

Titanium Dioxide - Uses, Applications, and Advances
Edited by Carlos Montalvo Romero, Claudia Aguilar and Edgar Moctezuma
p. cm.
Print ISBN 978-1-83769-455-6
Online ISBN 978-1-83769-454-9
eBook (PDF) ISBN 978-1-83769-456-3

Meet the editor

Dr. Carlos Montalvo Romero completed his undergraduate studies and Ph.D. in Chemical Engineering Sciences from the Autonomous University of San Luis Potosí in Mexico. He is a research professor at the Autonomous University of Carmen in Mexico, teaching subjects in the Chemical Engineering degree program. His administrative tasks include being a member of the university's editorial committee and the head of the Chemical Kinetics and Photocatalysis Laboratory. He has a national patent registered at the IMPI (MX/A/2008/011793) and another in process (MX/A/2018/004811). He belongs to the National System of Researchers Level I. He is currently directing a master's thesis in reactor design and catalyst synthesis to obtain hydrogen, ultrasound treatments, and treatments of pollutants. He has authored seven book chapters and edited a book with more than 485 citations since 2018.

Claudia Aguilar holds a Doctorate in Chemical Engineering from the Autonomous University of San Luis Potosí (UASLP). She is a research professor at the Autonomous University of Carmen and a National System of Researchers member. Her research focuses on developing new catalytic materials with environmental applications, chemical reaction engineering, and catalytic reactor design. She serves as an evaluator for the National Council of Science and Technology (CONAHCYT) in the RCEA Engineering and Industry area and has also evaluated Frontier Science Projects (2022-2023). In 2022, she was named a Distinguished Researcher by the National Association of Faculties and Schools of Chemistry (ANFEQUI). More recently, she received a distinction from the Government of the State of Campeche, Mexico, through the Institute of Women of the State of Campeche, in collaboration with COESICYDET (State Council of Science and Technology of Campeche).

Dr. Edgar Moctezuma graduated from Ohio State University in 1991 and has been a full-time professor at Universidad Autonoma de San Luis Potosí in Mexico since 1991. His research focuses on selective oxidation of lower alkanes, photocatalytic oxidation of organic chemical compounds, and complete oxidation of Volatile organic compounds (VOCs). His work experience includes being a Quality Control Engineer at Uniroyal S. A., a Research Engineer at Hyl S. A., a Process Engineer at IMMSA, and a Professor Assistant at Ohio State University. He has supervised 15 Ph.D. graduates and 28 Master of Science graduates. He has published 90 papers in international journals (in English) and contributed to over 300 publications at international conferences. A member of the National System of Researchers (SNI) since July 1, 1994, he currently holds Level III status (CVU 13492). In December 2014, he received the University Award for Socio-Humanistic, Scientific, and Technological Research from the Autonomous University of San Luis Potosí (UASLP).

Contents

Preface

Titanium dioxide (TiO_2) has emerged in recent decades as a highly versatile material whose unique properties have placed it at the forefront of scientific and technological research. Its ability to interact with light, its chemical stability, and its low cost make it an ideal candidate for a wide range of applications, from the pigment industry to the treatment of contaminated water.

This book delves into the fascinating world of titanium dioxide, thoroughly exploring its properties, preparation methods, and most promising applications. Within its pages, the reader will find a detailed description of the mechanisms underlying heterogeneous photocatalysis, a phenomenon in which TiO_2, when irradiated with ultraviolet light, can generate highly reactive species that can decompose a wide variety of organic and inorganic contaminants present in water.

In addition, the optical properties of titanium dioxide will be discussed, making it an essential material in the manufacture of high-purity white pigments and the development of photovoltaic devices. The latest advances in synthesizing new TiO_2-based materials, such as nanostructures and doped composites, will also be discussed, which offer great potential for improving the efficiency of photocatalytic processes.

This book is aimed at students, researchers, and professionals interested in chemistry, engineering, and environmental sciences. We hope the book provides an updated and comprehensive overview of titanium dioxide and stimulates research in this constantly evolving field.

Carlos Montalvo Romero and Claudia Aguilar
Faculty of Chemistry,
Autonomous University of Carmen,
City of Carmen, Campeche, Mexico

Edgar Moctezuma
Faculty of Chemistry,
Autonomous University of San Luis Potosi,
San Luis Potosi, Mexico

Chapter 1

Perspective Chapter: Titanium Oxide – Uses and Applications

Leskey Mduduzi Cele

Abstract

This chapter discusses the uses and application of TiO_2 in industry covering the more well-known including uses in foods and catalysis as well as in construction but focusing more attention on energy storage and Li-ion cells, photocatalysis, sonodynamic therapy, and medical treatment of cancer and in diagnostic sensors. The conclusion drawn is that TiO_2 remains a widely used material although concerns remain about the possibility of toxicity and cytotoxicity and ultimate fate in the environment.

Keywords: titanium oxide, anatase, rutile, photocatalysis, drug delivery, heterogeneous catalysis

1. Introduction

TiO_2 is a ubiquitous metal oxide found in many consumer products and used widely in various industries. It is found in many cosmetic products including toothpaste, deodorants, sunscreens, face creams, lip balms, shampoos and shaving cream and even in foods such as sweets, coconut curd, donuts, gum, chocolate, white icing and sugar toppings [1]. Some of its most important uses are a consequence of its ability to serve as a transparent barrier to UV light and as a whitening agent in the products to which it is added.

According to Ayorinde et al. [2], 98% of TiO_2 is pigmentary and involve the use of particle sizes above 250 nm mainly pigmentary, while nanoparticles smaller than 250 nm generally have properties unique to those of bulk TiO_2 and are used in fields including catalysis, photocatalysis, nanomedicine, batteries and solar cells as well as in sensors and biosensors. It is estimated that 10,000 tons of nanoTiO_2 are produced annually.

TiO_2 particles come in various shapes including spheres, spheroids, wires and nanotubes and it has been demonstrated that many properties are size- and shape-dependent.

Concerns have been raised about the possible toxicity and cytotoxicity of TiO_2 and these have been addressed by many researchers. TiO_2 nanoparticles can penetrate biological barriers and make their way through or into skin, gastrointestinal tract, kidneys, brain and lung deposition, which has also been confirmed. Setyawati et al. [3] showed that titanium dioxide nanoparticles in the lower size range ($\sim$25 nm) are implicated in endothelial cell leakiness (ECL) even when they are not excited by light, while microcrystalline TiO_2 particles (~650 nm) do not disrupt the endothelial cell layer. Endothelial cell leakiness was found to be a consequence of the disruption interaction of endothelial cells with TiO_2 nanoparticles and to be implicated in tumour cell metastasis and chronic vascular disease. Other studies also indicate that very small bare anatase nanoparticles (10 and 20 nm) can induce DNA damage and micronuclei formation in a human bronchial epithelial cell line in the absence of photoactivation. While the above-discussed are of extreme concern and suggest the necessity for careful assessment of the risks associated with exposure to TiO_2, it should be noted that other results emphasise that human studies have not been able to detect any relation between TiO_2 occupational exposure and the cancer risk and that the toxicity of TiO_2 particles at concentrations below 1 mg/mL has not been proved. This makes the current status of our knowledge inconclusive [4–6].

Rajh et al. [4] and Irshad et al. [5] report on two approaches in the industrial synthesis of TiO_2. In the older sulphate process, lower-grade ores are used in a batch process to synthesise mainly anatase and rutile in a process following the steps below:

$$TiOSO_4(aq) + 2H_2SO_4 \rightarrow TiOSO_4 + FeSO_4\,(aq) + 2\,H_2O. \quad (1)$$

$$TiSO_4 + H_2O \rightarrow TiO_2(s) + H_2SO_4. \quad (2)$$

$$TiO_2.nH_2O(s) \rightarrow TiO_2(s) + nH_2O. \quad (3)$$

The process is characterised by a large amount of waste and the requirement for expensive pollution control measure.

The relatively newer chloride process produces only rutile and requires higher-grade ores. It is a continuous process resulting in small amounts of waste and the possibility.

$$TiCl_4(aq) + O_2 \rightarrow TiO_2(s) + 2Cl_2. \quad (4)$$

Other smaller-scale synthesis routes include both physical and chemical methods. Physical methods include gas phase condensation, metal alloying, thermal crystallisation and molecular beam epitaxy (MBE) [5]. Chemical methods include sol-gel technique, hydrothermal process, chemical precipitation and solid-state metathesis [7].

This brief review of the uses and applications of TiO_2 is not exhaustive but is meant to merely provide examples of the broad use of TiO_2 in many sectors of human activity. The more widely known traditional uses of TiO_2 are covered in passing while more attention will be paid to those that are still in the earlier stages of development and likely to be less known. This includes uses in photocatalysis, degradation of recalcitrant organic molecules, energy storage devices and medical applications.

2. Structure general properties and chemistry

The crystal structural, optical and electrical properties of rutile, anatase and brookite are summarised in **Table 1** below (**Figure 1**) [8, 9].

The optical gaps for the three forms are all slightly above 3 eV with rutile ~3.0 eV, anatase ~3.4 eV and brookite ~3.3 eV [9], and as a result, natural TiO_2 is only photoactive in the UV region of the electromagnetic spectrum and not particularly

Properties	Rutile	Anatase	Brookite
Crystal structure	Tetragonal	Tetragonal	Orthorhombic
Lattice constant (Å)	a = 4.5936 c = 2.9587	a = 3.784 c = 9.515	a = 9.184 b = 5.447 c = 5.154
Molecule (cell)	2	2	4
Molecular volume ($Å^3$)	3121	34.061	32.172
Density ($g\ cm^{-3}$)	4.13	3.79	3.99
Ti-O bond length (Å)	1.949–1.980	1.937–1.965	1.87
Ti-O bond angle	81.2^0–90^0	77.7^0–92.6^0	77.0^0–105^0
Band gap at 10 K	3.051 eV	3.46 eV	
Conductivity	n-type semiconductor		

Table 1.
Crystal structural, optical and electrical properties of TiO_2 crystal structures.

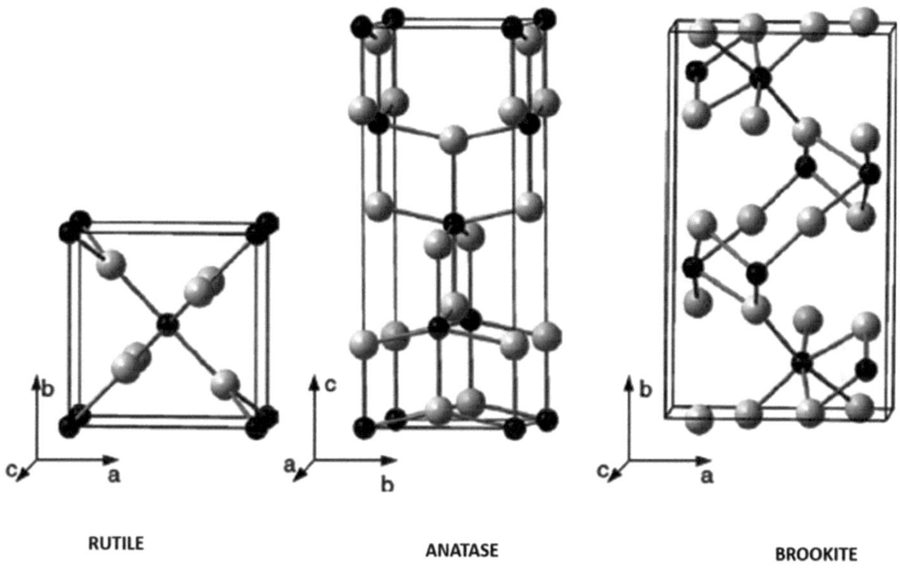

Figure 1.
The crystal structures of anatase, brookite and rutile (Stanford Advanced Materials, 2024).

efficient as an active solar cell material. The same property makes TiO_2 more effective as a skin barrier against UV radiation.

It is well established that the optical, chemical and electronic properties of TiO_2 particles are significantly influenced by their size and shape. Generally, particles larger than 250 nm are used mainly in relation to food and pigments (constituting roughly 98% of TiO_2 use) and those lower than 250 nm account for the rest. Furthermore, size reduction below 20 nm has been found to cause surfaces of metal oxide materials including TiO_2 to experience rearrangement of atom positions compared to bulk materials. The coordination sphere on the surface is incomplete and surface sites participate in atom trapping. Surface atoms are coordinated with solvent molecules resulting in weaker covalent bonds. This moves the energy levels of surface atoms to the mid-gap region and reducing their reducing and oxidising ability. Surface Ti atoms also adjust coordination environment in the nano-range resulting in compression of TiO bonds and extension of Ti-Ti bonds.

Rajh et al. [4] reported that size and shape have been successfully used to control electronic and chemical properties. Both influence particle crystallinity that makes it possible to control electron-hole transport and charge separation and regulate redox properties. They also determine the exposure of crystal planes that control selectivity and efficiency of photocatalytic properties.

Many of the most interesting properties and applications of TiO_2 are a consequence of its ability to produce OH^- and O_2^- radicals, and positive holes when irradiated with light as shown below (**Figure 2**).

According to Rajh et al. [4], the use of TiO_2 in environmental applications has its basis in the following reactions:

$$\mathrm{TiO_2} \xrightarrow{hv} \mathrm{e^-_{cb}} + \mathrm{h^+_{vb}} \tag{5}$$

$$H_2O + h^+_{vb} \rightarrow OH^\bullet + H^+ \tag{6}$$

$$O_2 + e^-_{cb} \rightarrow O_2^{-\bullet} \tag{7}$$

$$O_2^- + H^+ \rightarrow HO_2^\bullet \tag{8}$$

$$O_2^{-\bullet} + h^+_{vb} \rightarrow {}^1O_2 \tag{9}$$

$$2HO_2^\bullet \rightarrow H_2O_2 + O_2 \tag{10}$$

$$H_2O_2 + O_2^{-\bullet} \rightarrow OH^\bullet + OH^- + O_2 \tag{11}$$

$$h^+_{vb} + 2O_2^{-\bullet} \rightarrow O^\bullet_{ad} \tag{12}$$

The reactions above show the formation of positive holes and promotion of valence bond electrons to the conduction band. These, in turn, react with water molecules to

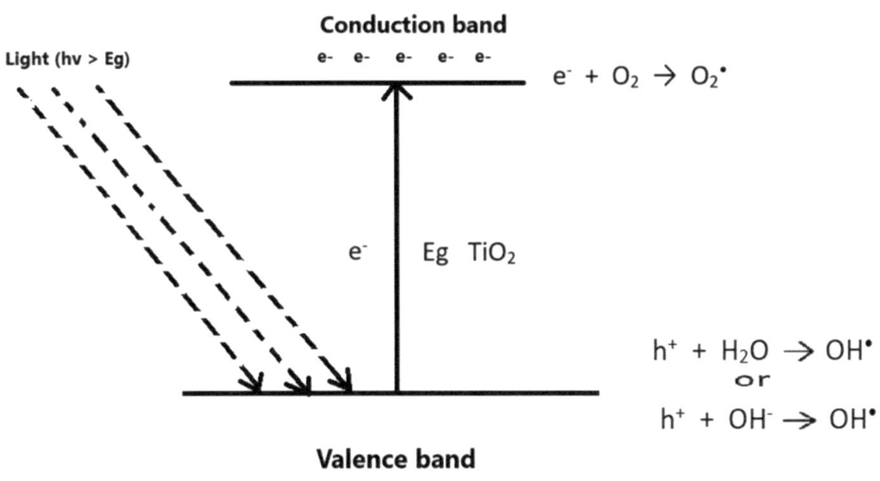

Figure 2.
The mechanism for the generation of OH and O_2* radicals from TiO_2 under irradiation by UV light [9].*

produce various radicals including $OH^{\cdot}$, $O_2^{\cdot}$ and $O_2^{-\cdot}$ which are capable of decomposing many organic species.

3. Applications

3.1 Cosmetics

TiO_2 owes much of its use in pigments to the fact that it does not absorb in the visible region of the electromagnetic spectrum (380–700 nm) and therefore appears white to the eye. In terms of reflection, the amount of reflected light is given by the Eq.

$$R = \frac{(n_P - n_m)^2}{(n_P + n_m)^{3.}} \tag{13}$$

where R is the amount of light reflected and n_P and n_m are the refractive indices of TiO_2 and the medium, respectively [8]. The equation shows that the amount of reflected light increases as the difference in the refractive indices increases. TiO_2 has a refractive index of 3.6–4.0 and is used in sunscreens where it is well known for its natural opaqueness, effective UV-blocking efficiency and the fact that it causes no skin irritation [10]. In addition to that, it is commonly found in cosmetic products including toothpaste, deodorants, sunscreens, face creams, lip balms, shampoos and shaving cream where its main role is to enhance the white colour [11].

In TiO_2-ZnO composites where its concentration is as low as only 5%, TiO_2 has been observed to increase the refractive index by a factor of 7. TiO_2 absorbs UVB, while ZnO absorbs more UVA-1. The explanation is that the TiO_2 valence band has many densely packed electronic states that allow many adsorption possibilities, while for ZnO below 100 nm, electronic energy levels behave like discrete energy levels resulting in a blue shift and less opaqueness [12].

3.2 TiO_2 in photocatalysis

Solar energy is viewed as a better alternative to fossil-based fuels since it is clean, abundant and inexhaustible [11]. New semi-conducting materials are increasingly required for use in solar cells and nano-TiO_2 is increasingly viewed as an important material of the future in this regard [13–19]. Indeed, a lot of work on improving and refining TiO_2 for this purpose is proceeding and showing promise. The use of TiO_2 includes its incorporation into dye-Sensitised solar cells (DSSCs) and, lately, perovskite solar cells (PSCs). In all similar applications, TiO_2 promotes device efficiency and improves operational stability.

Dye-sensitised solar cells convert solar energy to electric energy by light sensitization. They typically consist of a semiconducting metallic oxide such as TiO_2 is a setup similar to the one shown in **Figure 3** below. In addition to the two electrodes, the cell contains a dye-sensitised nanostructured TiO_2 mesoporous layer and a liquid electrolyte. The mesoporous TiO_2 layer with a monolayer of the dye at its surface is placed in contact with the redox electrolyte. When irradiated with solar energy, the dye injects electrons into the TiO_2 conduction band and the electrical power is produced by conducting these to the external circuit. The superior effectiveness of the TiO_2 mesoporous layer is a consequence of its large surface area [9].

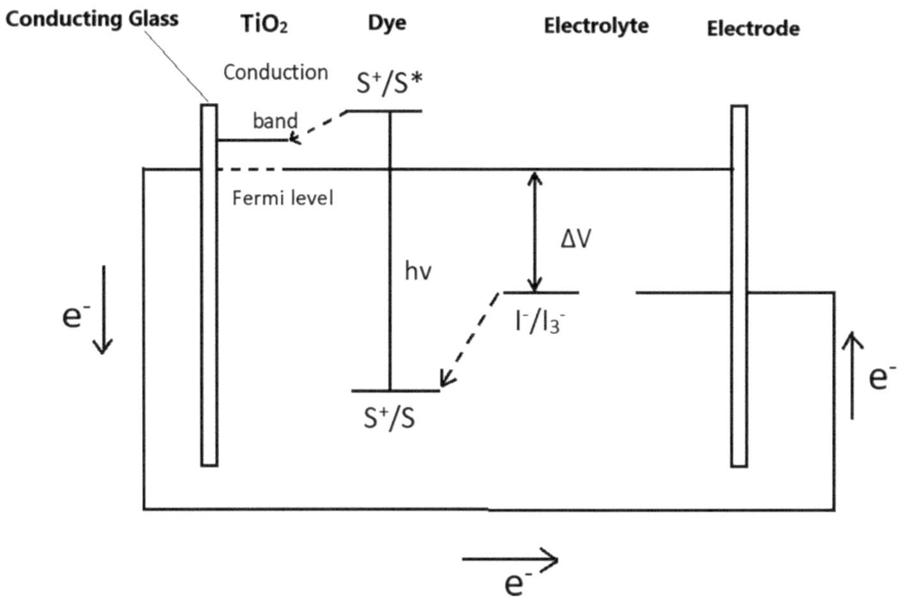

Figure 3.
Working principle of dye-sensitised solar cells [9].

Liu et al. [11] also reported extensively on TiO_2 in dye-sensitised solar cells. Their work shows the importance of particle size and indicates that brookite has very low Li-ion intercalation capacity above 30 nm size, while the rest of the TiO_2 polymorphs actually have active Li-ion storage capabilities. Compared to bulk TiO_2 that has a specific capacity of 160 $mAhg^{-1}$ (but drops below 50 $mAhg^{-1}$ during use) with intercalation level 0.5 Li per Ti, the specific capacity for nano-TiO_2 remains stable at 150 $mAhg^{-1}$ amounting to 1 Li per Ti. The reason is that there is a different phase transformation mechanism in operation for nano-TiO_2. For TiO_2 nanotube and nanowire arrays, the values exceed 456 $mAhg^{-1}$ [9].

TiO_2 has also proved useful in photocatalysis when used in conjunction with other nanoparticles and metallic oxides including those containing bismuth. In our own laboratory, Mahhumane et al. [20] used a ternary z-scheme heterojunction of Bi_2WO_6 with carbon nanoparticles (CNP) and TiO_2 nanotube arrays to remove paracetamol from water by photoelectrocatalysis. The results showed that at optimal conditions, the electrode was applied for photoelectrocatalytic degradation of paracetamol, which gave a degradation efficiency of 84% within 180 min. The total organic carbon removal percentage obtained when using this electrode was 72%. Scavenger studies revealed that the holes played a crucial role during the photoelectrocatalytic degradation of paracetamol. The electrode showed high stability and reusability therefore suggesting that the z-scheme Bi_2WO_6-CNP-TiO_2 nanotube array electrode is an efficient photoanode for the degradation of pharmaceuticals in wastewater.

Perovskite solar cells (PSCs) consist of a substrate made of transparent and conducting glass onto which an electron-sensitive or hole-sensitive contact is placed followed by a perovskite absorbing layer, another electron- or hole-selective layer and a metal contact. In these, TiO_2 functions as the mesoporous contact. Seo et al. [21] were able to demonstrate that the incorporation of TiO_2 into TiO_2-PSC resulted in significant improvement over PSC-only cells. In their work on PSCs, Correa-Baena et al. [22] also demonstrated that substances such as $CH_3NH_2PbI_3$ could be used to get electrons into the conduction band of TiO_2 to improve cell efficiency beyond 20%. Furthermore, perovskites ate tunable and have better tolerances for structural defects compared to silicon that needs to be of exceptionally high purity. The only major drawback is the low durability of perovskite solar cells which will have to improve significantly to compete with existing competing technologies (**Figures 4–8**).

As shown above, these results show that the TiO_2 nanocomposites were successful in the degradation of paracetamol and prove the fact that the electrodes are reusable.

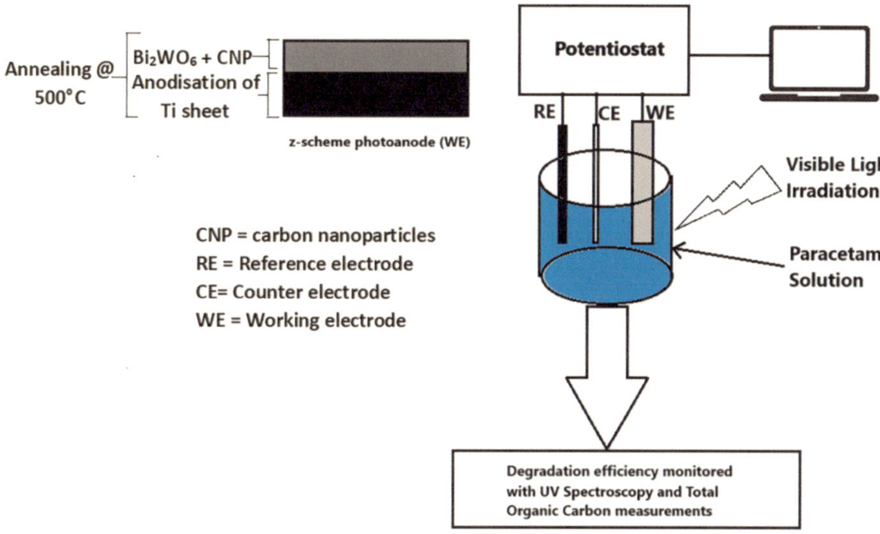

Figure 4.
Schematic diagram of prepared Bi_2WO_6-CNP-TiO_2 NTA z-scheme photoanode used for photoelectrocatalytic degradation of paracetamol in a three-electrode cell [20].

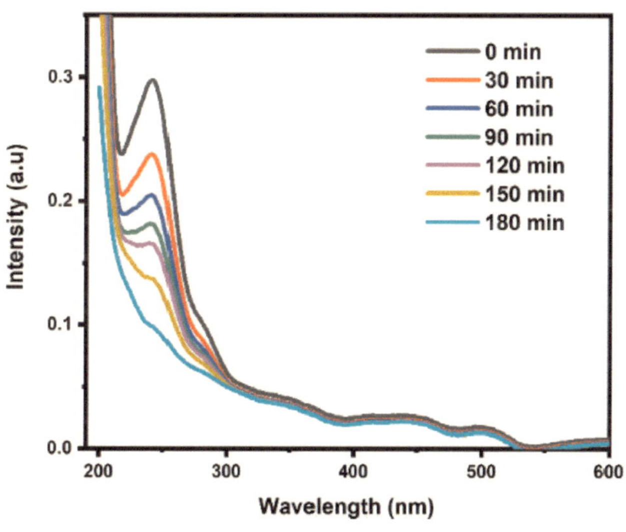

Figure 5.
UV–vis degradation of paracetamol [20].

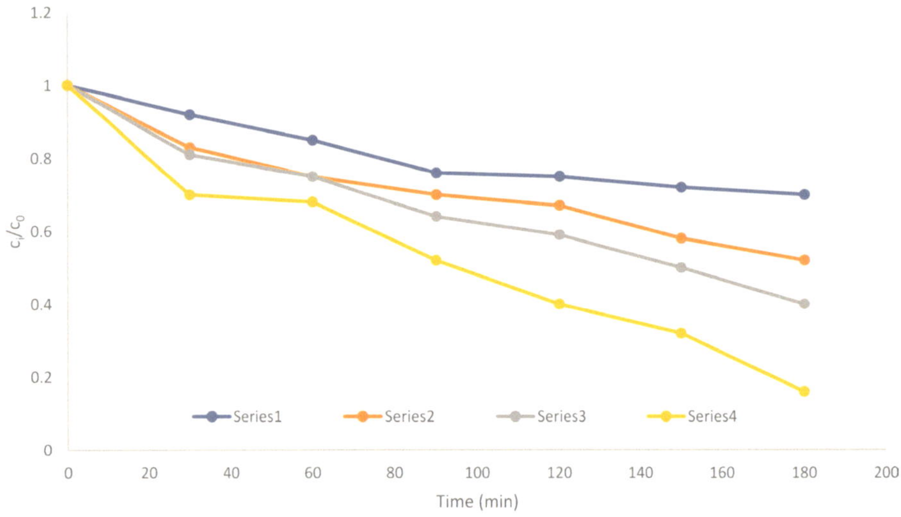

Figure 6.
Normalised concentration decay versus time plot for photocatalytic, electrocatalytic and photoelectrocatalytic degradation of paracetamol on of Bi_2WO_6-CNP–TiO_2 NTA electrode [20].

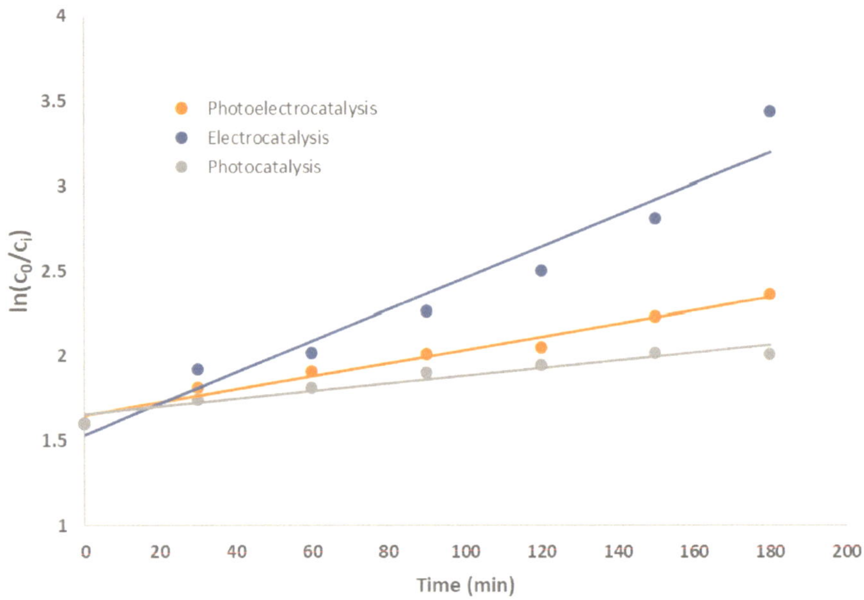

Figure 7.
Kinetics plots for photocatalytic, electrocatalytic and photoelectrocatalytic degradation of paracetamol on of Bi_2WO_6-CNP–TiO_2 NTA electrode [20].

The results indicate that the hole and hydroxyl radicals play a major role in the degradation of paracetamol, while superoxides play a minor role. The photoelectrocatalytic degradation of paracetamol may be considered to follow the equations below:

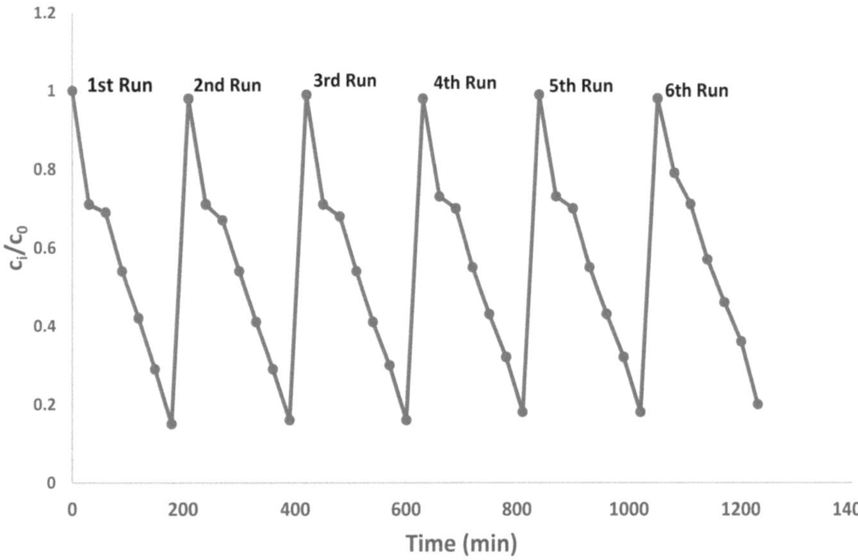

Figure 8.
Reusability cycle experiments for the degradation of paracetamol on of Bi_2WO_6-CNP–TiO_2 NTA electrode [20].

$$hv + Bi_2WO_6 - CNP - TiO_2\ NTA \rightarrow Bi_2WO_6 - CNP - TiO_2\ NTA\left(h_{VB}{}^{+} + e_{CB}{}^{-}\right) \quad (14)$$

$$e_{CB}{}^{-} + O_2 \rightarrow {}^{\bullet}O_2{}^{-} \quad (15)$$

$$h_{VB}{}^{+} + H_2O \rightarrow {}^{\bullet}OH + H^{+} \quad (16)$$

$${}^{\bullet}OH + \text{Paracetamol} \rightarrow CO_2 + H_2O \quad (17)$$

$${}^{\bullet}O_2{}^{-} + \text{Paracetamol} \rightarrow CO_2 + H_2O \quad (18)$$

$$h_{VB}{}^{+} + \text{Paracetamol} \rightarrow CO_2 + H_2O \quad (19)$$

The final example of the use of TiO_2 in photocatalysis is the use of TiO_2 nanoparticles in the photodegradation of methylene blue dye which revealed the unexpectedly higher degradation efficiency with biosynthesised TiO_2 nanoparticles when compared with their chemically synthesised counterparts [23]. This was also found to be true for antibacterial activity against gram-positive and gram-negative bacterial strains. This confirms to the important of the structural features of TiO_2 nanoparticles in their chemistry.

3.3 Production of hydrogen and Li-ion batteries

A more recently reported use of TiO_2 is in the photocatalytic production of hydrogen which occurs according to the mechanism depicted below (**Figure 9**).

The process starts with the excitation of electrons by photons of sufficient energy (e.g. exceeding the band gap) to the conduction band (CB) which creates holes (h^+) in the valence band of TiO_2. The holes and electrons can combine to release energy in the form of heat or photons but if they do not, they may migrate to the surface where they reduce and oxidise adsorbed water molecules to generate molecular hydrogen and oxygen.

For energy storage, the use of TiO_2 is exemplified by its use in Li-ion batteries. The reaction in operation is:

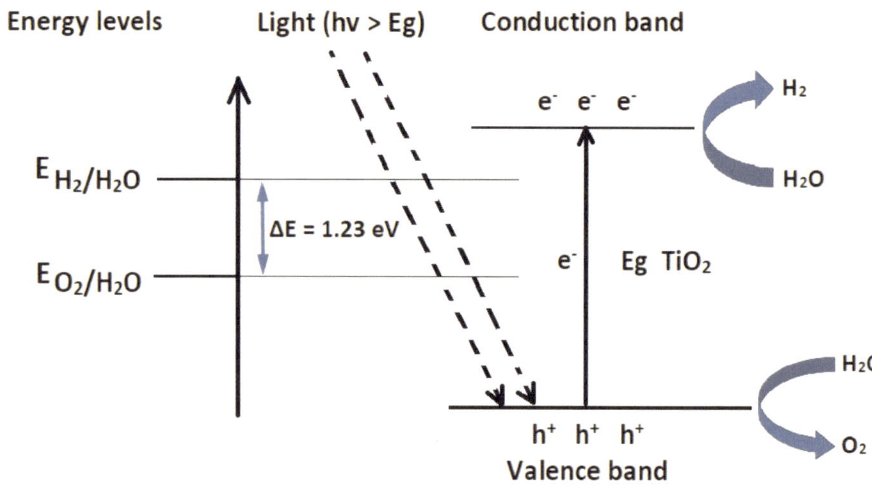

Figure 9.
The mechanism for the photocatalytic production of hydrogen by TiO_2 [9].

$$xLi + TiO_2 + xe^- \rightarrow LIxTiO_2, \text{where } 0 < x < 1 \quad (20)$$

and the specific capacity is typically in the range of 335 $mAhg^{-1}$ (Wu). The limitations for these are low electrical conductivity and poor rate capability [9].

In studies on the Li-ion intercalation capacity, Liu et al. [11] observed that brookite has a very low capacity once the nanoparticles go above 30 nm in size, while anatase and rutile have active Li-ion capability even above that. While bulk TiO_2 has a measured specific capacity of 160 $mAhg^{-1}$, corresponding to an intercalation level of 0.5 Li per Ti. For nano-TiO_2, values exceeding 300 $mAhg^{-1}$ have been reported, which correspond to intercalation levels of 1 Li per Ti. This is explained in terms of the different phase transformation mechanism in operation [11]. When fabricated by hydrolysis of titanium alkoxide and heat-treated above 250°C, nano-TiO_2 particles show specific capacities stable at 150 $mAhg^{-1}$. TiO_2 nanotube and nanowire arrays have yielded specific capacities of >356 $mAhg^{-1}$ [11].

3.4 Heterogeneous catalysis

In heterogeneous catalysis, the objective is to use the support to improve the stability of the active nanoparticles (usually metallic or bimetallic), optimise performance, reduce the amount of metal (which is usually expensive) and improve activity through metal-support interactions. TiO_2 has proven to be quite useful in heterogeneous catalysis because of its tunable porous surface mechanical strength and thermal stability [24–26]. The drawbacks include low surface area and its low adsorption ability.

Examples of the many uses of TiO_2 in heterogeneous catalysis are given in **Table 2** below.

- Use as supports for V_2O_5-WO_3/TiO_2 for selective catalytic reduction of NOx (excellent thermal stability and lower oxidant activity for conversion of SO_2 to SO_3) [12]:

$$NOx \rightarrow N_2 + H_2O \quad (21)$$

Catalyst	Reaction
Cu/TiO_2	Oxidation of CO
V_2O_5/TiO_2	Oxidation of o-xylene to phthalic anhydride Hydrogenation of propane Pollution abatement of NO
Pd/TiO_2	Oxidation of methanol Acetoxylation of toluene
$PdNi/TiO_2$	Oxidation of methanol
$CoMn/TiO_2$	Fischer-Tropsh synthesis
Au/TiO_2	Hydrosulphurization of hydrocarbon oils Selective reduction of NO
Co/TiO_2	CO conversion
Ni/TiO_2	Hydrogenation of maleic acid
Pt/TiO_2	Oxidation of acetaldehyde

Table 2.
TiO_2 in heterogeneous catalysis [22].

3.5 Construction

TiO_2 is used for the manufacturing of coatings capable of decomposing adsorbed organic under the action of sunlight and rain water. This makes it ideal for use in the production of self-cleaning windows and tiles that are now used widely in the construction industry. Furthermore, tiles containing TiO_2 have been demonstrated to be useful as photocatalytic surfaces for decompositions of bacteria and viruses. The addition of silver to form TiO_2-Ag results in enhanced photocatalytic properties [9]. In these construction materials, hydroxyl radicals decompose adsorbed organic substances on TiO_2 surface. Since TiO_2 is hydrophilic, the contaminates are easily washed off by rainwater. Pilkington glass has already successfully debuted to the marked self-cleaning windows that are coated with TiO_2 for photocatalysis and hydrophilicity which allow photocatalysis to destroy contaminants that are then washed off as thin water layers instead of droplets [9].

3.6 Medicine

3.6.1 Cancer treatment

In biomedical applications, the important properties of TiO_2 are its extraordinary stability, exceptional photo-reactivity and biocompatibility [27–35]. As has been already established, TiO_2 responds to light to produce reactive oxygen species (ROS) which are capable of altering cell functioning and can be directed to selectively target cancer cells [9].

Photodynamic therapy (PDT) and sonodynamic therapy (SDT) are two broad methods that have been employed in cancer treatment. Photodynamic therapy is an antitumor method in which a photosensitive agent is applied to the target area which is then illuminated for the activation of the therapeutic agent. In sonodynamic therapy, ultrasound irradiation is used to promote the production of $OH^{\bullet}$ radicals even in the absence of light. In this way, localised heating is generated or activation of drug release on site is achieved. The advantage of this method is that it is capable of acting deep below the skin surface and therefore allows the targeting of deeper lying tissues.

Here are some examples of the use of TiO_2 in ways that are advancing cancer treatment:

- HeLa cells were shown to be susceptible to exposure to UV light in the presence of TiO_2 [9].

- TiO_2-PEG (polyethylene glycol) on glioma cells. TiO_2-PEG could be a novel sensitiser for SDT because cell damage was found to be enhanced by TiO_2-PEG. Sonotoxicity was found by Yamaguchi et al. to be almost proportional to the duration of ultrasound (US) exposure.

- Catalysis of oxidative DNA damage in human fibroblasts.

- Chlorine e6 and hypocrellin B used in visible-light-induced cell death in which the significant damage in thymic lymphoma cells was demonstrated.

- Zinc phthalocyanine molecules in TiO_2 porous networks shown to be active against tumour cells [4].

A major limitation in the use of TiO_2 is its poor absorption in the visible range. In this regard, effective ways of extending the optical absorption of TiO_2 into the visible region have been found and they include the following:

- The synthesis of TiO_2-enediol composites by combining TiO_2 with enediol molecules, which facilitates the repair of uncoordinated Ti surface atoms to restore the six-coordinated octahedral geometry. The excitation of electrons from chelating enediol ligands lowers the TiO_2 band gap.

- Dye-sensitisation of TiO_2 nanoparticles entails the modification of TiO_2 nanoparticles with a monolayer of chlorin e6 trisodium salt followed by irradiation with visible light in the range 550–750 nm. This results in improved effectiveness in this range, which has been demonstrated by Rajh et al. [4] to enable significant damage to EL-4 cells. The dye-conjugate TiO_2 particles are demonstrably more effective than the dyes alone, and this confirms that a cooperative interaction between the dye and TiO_2 particles is in play.

- Sensitisation by anionic and cationic doping and by doping with nitrogen to produce N-TiO_2 has shown enhanced absorbance up to 600 nm. Visible-light induced photolysis results in induced cytotoxicity, which has been proven for hepatocellular carcinoma cells, HeLa cells and nasopharyngeal carcinoma cells.

- Surface modification with inorganic complexes to produce composites such as TiO_2-O-$PtCl_4(H_2O)$ has been proved to result in phototoxic effects when compared with TiO_2 against mouse melanoma cells.

Bogdan et al. [34] have written extensively on the use of electroporation as part of cancer treatment. Electroporation is a physical method of introducing molecules into cells as part of therapy. In their work, electroporation resulted in improved selectivity and effectiveness of photoexcitation of cancer cells which, in turn, improved the delivery of the agent to the target area. Xu et al. [36] also demonstrated the high efficiency of antibody-TiO_2 composites (Ab-TiO_2) in killing cancer cells with even low concentrations of Ab-TiO_2. This makes it possible to use this method against a wide range of cancer cells without changing the antibody used.

All in all, the following three steps are accepted to be crucial in all oxidative stress tumour therapies:

1. Increased levels of ROS and oxidative products as a result of excitation of a suitable agent such as TiO_2 or TiO_2-composite.

2. ROS reaction with lipids in cellular membranes and nucleotides in proteins and DNA.

3. Cell death.

The importance of shape and size is crucial since work on rat cells has shown a strong correlation between the TiO_2 diameter and its genotoxicity [37–39].

Photodynamic therapy and sonodynamic therapy are thought to be the future of cancer therapy and believed to be aiding the progress toward less invasive and more site-specific treatment with less side effects. The remaining difficulties include difficulties in the preparation of drug formulas that allow targeted delivery on TiO_2 nanoparticles to target cancer cells and this has, so far, restricted the application of both methods.

3.6.2 Drug delivery

The main goal of ongoing work on drug delivery systems is to improve controlled time release and site selectivity. Shape and size-selected TiO_2 nanoparticles aid in control of electronic and chemical properties. This is achieved by using light to activate the action of TiO_2 in what is referred to as light-induced drug delivery.

Examples of TiO_2 in drug delivery:

- Binding of a protein to TiO_2 to target the TiO_2-protein conjugate to a specific cell protein or organelle. In this manner, monoclonal antibodies have been linked TiO_2 semiconducting nanoparticles, which are able to bind to cells and alter cell functioning through redox interactions. Antibodies linked to TiO_2 promise to provide future platforms for targeting therapy such as cancer treatment [4].

- TiO_2 linked to T-cell-specific antibodies irradiated with UV-vis showing light-induced redox chemistry. This has been shown to result in the light-induced death of targeted T-cells without affecting healthy cells. Definite binding to targeted cells has been confirmed [4].

- Rozhkova et al. [27] studied glioblastoma multiform (GBM) cells targeted with TiO_2-multiclonal Abs and confirmed concentration-dependent phototoxicity toward targeted cells. Furthermore, this toxicity was proved to be mediated by superoxide radical anions. It was further observed that the superoxide triggers call apoptosis, while singlet oxygen causes photosensitised damage to plasma membranes.

- The use of polylactic coglycolic acid (PLGA) /TiO_2 advanced 3D-printed systems and nanosystems for drug delivery and tissue engineering applications by Rasoulianboroujeni et al. [28].

Two broad approaches are recognised in the use of TiO_2 in drug delivery:

1. Use of TiO_2 nanotubes containing hydrophilic drugs surface-modified with hydrophobic ligands to control release [29].

2. Use of physisorption to attach oxygen-rich drug on TiO_2 whiskers. Integration of drugs with TIO_2 resulted in altered uptake mechanism [30].

3.6.3 Imaging-guided therapy

In image-guided therapy (IGT), the therapeutic effect is monitored and the therapeutical accuracy improved by imaging the affected area to visually locate the therapeutic agent in a living system. This makes it possible to regulate the treatment plan in time and reduce the damage to health tissues [31].

Examples that illustrate this include the coupling TiO_2 with dyes such as alizarin red results in the formation of a complex that has different optical properties to the free dye and fluoresces red to show accumulation of the complex within the targeted area and allows imaging that shows the expected accumulation [23] and the imaging of Flavin-TiO_2 composites localised in the cytoplasm of BT-20 cancer cells [24].

3.6.4 Sonodynamic therapy (SDT)

Ninomiya et al. [37] and others have reported on the use of protein-modified TiO_2 nanoparticles in sonodynamic therapy. This uses ultrasound irradiation to enhance OH radical generation resulting in the degradation of some substances and inactivation of microorganisms. The HEPG2 cell line has been targeted successfully in this way and it was found that the effect of ultrasound is an important factor in the effectiveness of the process.

In other studies, OH^* and hydroperoxyl (HO_2^*) radicals are produced by illumination from H_2O in the presence of TiO_2. Also, oxidation of cells is by photogenerated holes. TiO_2 particles have been observed localised on cell membranes and cytoplasm.

3.7 Sensors

Electrical biosensors are used for drug discovery, diagnostics, environmental applications and food safety. High-surface area electrodes significantly increase signal-to-noise ratio in electrochemical detection of metabolically relevant redox proteins or tumour markers.

Examples on the use of TiO_2 in biosensors:

- The report by Wang et al. [35] proved the detection of NADH with dopamine-modified TiO_2 nanocrystals films excited by VIS light, while Lin et al. reported on the use of TiO_2 composites in the detection of IgG secondary antibodies.

- The use of 3-aminopropyl bimetoxysilane (APTMS) and glutaraldehyde coupling linkers to link CDH peptides to TiO_2 films for the detection of HIV-1 viruses [40]. Importantly, other studies also show that DNA mismatches can be detected using charge separation in TiO_2 nanoparticles. During DNA sensing, DNA mismatches occur when template (probe) DNA is bound to target DNA with an altered base pair [4].

3.8 Other applications and fate in soil

TiO_2 is found in many common edible items including sweets, coconut curd, donuts, gum, chocolate, white icing and sugar toppings where it is mainly used as a whitening agent [1]. Further uses include in the manufacturing of high-end paper where it improves whiteness, increases density, smoothness and strength but also causes high abrasion and increases costs. It is also used in coatings, printing inks, chemical fibres, rubber, fabrics and textiles where its role is mainly as a white pigment [24].

The ultimate fate of TiO_2 in soil is a matter of concern, which requires further study. Weir et al. [1] and Thiagarajan [10] found that this is affected by NOM, clay content, microbial activity, ionic strength and pH.

3.9 TiO_2 nanotubes, nanowires and nanoribbons: Synthesis and applications

The synthesis of fullerenes, carbon nanotubes and carbon nanorods in the mid- to late 80s and early 90s opened up interesting possibilities in the synthesis and use of TiO_2 polymorphs other than the long-known particles and nanoparticles. In particular, the arrival of TiO_2 nanotubes, nanowires and nanoribbons has opened up possibilities beyond what could be envisaged in the recent past.

Among many others, Yuan and Su [41] and other researchers [42–47] reported on titania nanotubes and, generally, the nanotubes were well crystallised with a shell spacing of 0.78 nm and ranged from 10 to 15 nm in diameter 1–5 layers.

Various synthesis methods have been reported and, predictably, the properties of the resulting nanotubes differ in many structural and other properties. Yuan and Su [41] synthesised titania nanotubes by hydrolysis of triisopropoxide in ethanol followed by calcination. They obtained yields of 80–90% and observed that in some cases, the number of layers, as viewed under TEM, was unequal, which suggests that the nanotubes are formed from rolled up sheets of TiO_2 and not concentric as generally observed with carbon nanotubes. In the hydrothermal treatment process, they also reported that the ultimate product depends on the source of titania used. While titania nanotubes were the major when crystalline rutile and anatase were used in the temperature range 100–150°C, amorphous TiO_2 produced no nanotubes but mainly non-tubular, needle-shaped fibres in the temperature range 100–160°C, nanoribbons between 180 and 250°C and nanowires when KOH was used instead of NaOH during hydrothermal treatment.

This was confirmed by Kustiningsih et al. [48] who synthesised nanotubes with surface areas of up to 123 m^2g^{-1} with NaOH and nanowires with surface areas up to

115 m^2g^{-1} when KOH was used instead. In all cases, no change in the band gap energy was measured but better photocatalytic activity for hydrogen production was observed with TiO_2 nanotubes presumably because of more effective contact between the active sites and water molecules, which improves the photocatalytic activity.

Lim and Choi [49] and Tan et al. [50] demonstrated the synthesis of anodically grown titania nanotubes through anodic oxidation of titanium in ethylene glycol. Tan et al. [50] went further to compare different synthetic methods and experimented with the assisted template method, anodic oxidation and hydrothermal treatment. They found that the assisted template method produced large nanotubes with uniform sizes and achieved dimensional control by varying the properties of the porous alumina templates used. Hydrothermal treatment produced higher purity nanotubes with good crystallinity and dimensional control was possible through control of synthesis parameters such as pH and synthesis time. Anodic oxidation produced amorphous, aligned nanotubes with the high aspect ratios. In terms of applications, they were able to grow osteoblasts into the nanotubes and found that adhesion was improved as a result of the filopodia of osteoblasts growing into the nanotube pores. Successes on the use of these nanotube on dental implants, pacemakers, knee and hip joints, bone plates and heart valves have been confirmed.

Lee et al. [51] were able to use TiO_2 nanotubes for applications in dye-sensitised solar cells (DSSCs), photocatalysis, sensors and Li-ion batteries. They also confirmed control of tube dimensions through control of anodisation parameters.

Wang et al. [47] reported applications in photovoltaics, DSSCs, solid-state sensors and energy storage through LI-ion batteries and supercapacitors. They found that the band gap increases with decreasing nanotube diameter as a result of quantum confinement effects and that TiO_2 nanowires have better electron diffusion coefficients as a result of reduced grain boundaries, which eliminates electron trapping and avoids charge recombination. This results in between DSSC performances. In fact, their results demonstrated that TiO_2 nanowires were far superior to TiO_2 nanoparticles in terms of DSSC performance. Finally, in work on gas sensors for water, NH_3, NO_2 and ethanol, they reported better sensitivity when TiO_2 nanotubes and nanowires were used instead of TiO_2 nanoparticles.

4. Conclusion

TiO_2 is widely used in many sectors of human activity and innovative applications are coming to the fore as more is discovered about the capabilities of the various forms especially newly discovered forms such as nanoparticles, nanotubes and nanowires. The size and shape of these is of crucial importance in the optical and electronic properties as well as their chemistry. Although concerns have been raised concerning human exposure, risk assessment has produced inconclusive results. There is no doubt that more applications will result from current work and that continued refinement of present applications is likely to result in improved efficiency and efficacy.

While it is expected that the more traditional uses of TiO_2 will continue to expand, the arrival to the scene of other one-dimensional polymorphs including nanotubes, nanowires and nanoribbons has opened up new possibilities. It has been demonstrated that these one-dimensional structures perform better in many applications and the fact that their properties may be tuned to specific applications by varying the synthesis parameters is of particular importance. This makes it possible to synthesise and modify materials in such a way that surface areas, band gaps and other properties are predictable and suited to specific requirements.

Judging from current trends and ongoing work, it is clear that TiO_2 in its various forms will continue to be an important compound in the development and applications of new technologies.

Author details

Leskey Mduduzi Cele
Tshwane University of Technology, South Africa

*Address all correspondence to: celelm@tut.ac.za

References

[1] Weir A, Westerhoff P, Fabricius L, Hristovski K, et al. Titanium dioxide nanoparticles in food and personal care products. Environmental Science & Technology. 2012;**46**(4):2242-2250

[2] Ayorinde T, Sayes CM. An updated review of industrially relevant titanium dioxide and its environmental health effects. Journal of Hazardous Materials. 2023;**4**:100085-100092

[3] Setyawati MI, Tay CY, Chia SL, Goh SL, et al. Titanium dioxide nanomaterials cause endothelial cell leakiness by disrupting the homophilic interaction of VE–cadherin. Nature Communications. 2013;**4**:1673-1685

[4] Rajh T, Dimitrijevic NM, Bissonnette M, Koritarov T, et al. Titanium dioxide in the service of the biomedical revolution. Chemical Reviews. 2014;**114**:10177-10216

[5] Irshad MA, Nawaz R, Rehman MZ, Adrees M. Synthesis, characterization and advanced sustainable applications of titanium dioxide nanoparticles: A review. Ecotoxicology and Environmental Safety. 2021;**212**: 111978-111992

[6] Kumar DK, Reddy KR, Sadhu V, Shetti NP, et al. Metal oxide-based nanosensors for healthcare and environmental applications. Nanomaterials in Diagnostic Tools and Devices. 2020:113-129

[7] Landmann M, Rauls E, Schmidt G. The electronic structure and optical response of rutile, anatase and brookite TiO_2. Journal of Physics: Condensed Matter. 2012;**24**:195503-195509

[8] Ali I, Suhail M, Alothman ZA, Alwarthan A. Recent advances in syntheses, properties and applications of TiO_2 nanostructures. RSC Advances. 2018;**8**:30125-30148

[9] Wu X. Applications of titanium dioxide materials. In: Titanium Dioxide: Advances and Applications. IntechOpen; 2022

[10] Thiagarajan V, Ramasubbu S. Fate and behaviour of TiO_2 nanoparticles in the soil: Their impact on staple food crops. Water, Air, and Soil Pollution. 2021;**232**:274-292

[11] Liu DW, Cao GZ, Wang Y. Positive Electrode: Nanostructured Transition Metal Oxides. Seattle, USA: Elsevier; 2009

[12] Smijs TG, Pavel S. Titanium dioxide and zinc oxide nanoparticles in sunscreens: Focus on their safety and effectiveness. Nanotechnology, Science and Applications. 2011;**4**:95-112

[13] Clemente A, Ramsden JJ, Wright A, Iza F. Staphylococcus aureus resists UVA at low irradiance but succumbs in the presence of TiO_2 photocatalytic coatings. Journal of Photochemistry and Photobiology B: Biology. 2019;**193**: 131-139

[14] Du J, Wang C, Zhao Z, Cui F, et al. Role of oxygen and superoxide radicals in promoting H_2O_2 production during VUV/UV radiation of water. Chemical Engineering Science. 2021;**241**: 116683-116690

[15] Haider AJ, Jameel ZN, Al-Hussain IHM. Review on titanium dioxide applications. Energy Procedia. 2019;**157**: 17-29

[16] Kleiman A, Meichtry JM, Vega D, Litter MI, et al. Photocatalytic activity of

TiO2 films prepared by cathodic arc deposition. Surface and Coatings Technology. 2020;**382**:125134-125162

[17] Pokhum C, Viboonratanasri D, Chawengkijwanich C. New insight into the disinfection mechanism of *Fusarium monoliforme* and aspergillus niger by TiO_2 photocatalyst under low intensity UVA light. Journal of Photochemistry & Photobiology, B: Biology. 2017;**176**:17-24

[18] Umer M, Tahira M, Azama MU, Tasleem S, et al. Synergistic effects of single/multi-walls carbon nanotubes in TiO_2 and process optimization using response surface methodology for photo-catalytic H_2 evolution. Journal of Environmental Chemical Engineering. 2019;7(5):103361-103369

[19] Zhao Y, Li Y, Sun L. Recent advances in photocatalytic decomposition of water and pollutants for sustainable application. Chemosphere. 2021;**726**: 130201-130216

[20] Mahhumane N, Cele LM, Muzenda C, Nkwachukwu OV, et al. Enhanced visible light-driven photoelectrocatalytic degradation of paracetamol at a ternary z-scheme heterojunction of Bi_2WO_6 with carbon nanoparticles and TiO_2 nanotube arrays electrode. Nanomaterials. 2022;**12**: 2467-2482

[21] Seo S, Shin S, Kim E, Jeong S, et al. Amorphous TiO_2 coatings stabilize perovskite solar cells. ACS Energy Letters. 2021;**6**:3332-3341

[22] Correa-Baena JP, Saliba M, Bounassisi T, Gratzel M. Promises and challenges of perovskite solar cells. Science. 2017;**358**(6364):739-744

[23] Aravind M, Amalanathan M, Mary MSM. Synthesis of TiO_2 nanoparticles by chemical and green synthesis methods and their multifaceted properties. Applied Sciences. 2021;**3**:409-419

[24] Mishra A. Analysis of titanium oxide and its applications in industry. International Journal of Mechanical Engineering and Robotics Research. 2014;**3**(3):561-566

[25] Bagheri S, Julkapli NM, Hamid SBA. Titanium dioxide as a catalyst support in heterogeneous catalysis. The Scientific World Journal. 2014;**12**:727496-727517. Available from: https://www.researchgate.net/publication/268228301 [Accessed: 15 October 15, 2024]

[26] Ramsden JJ, Sokolov IJ, Malik DJ. Questioning the catalytic effect of Ni nanoparticles on CO_2 hydration and the very need of such catalysis for CO_2 capture by mineralization from aqueous solution. Chemical Engineering Science. 2018;**175**:162-167

[27] Rozhkova EA. Nanoscale materials for tackling brain cancer: Recent progress and outlook. Advanced Materials. 2011;**23**:H136-H150

[28] Rasoulianboroujeni M, Fahimipour F, Shah P, Khoshroo K, et al. Development of 3D printed PLGA/TiO_2 nanocomposite scaffolds for bone tissue engineering applications. Materials Science & Engineering. C, Materials for Biological Applications. 2018;**23**: 96-105

[29] Song YY, Schmidt-Stein F, Bauer S, Schmuki P. Amphiphilic TiO_2 nanotube arrays: An actively controllable drug delivery system. Journal of the American Chemical Society. 2009;**131**:4230-4241

[30] Li Q, Wang X, Lu X, Tian H, et al. The incorporation of daunorubicin in cancer cells through the use of titanium

dioxide whiskers. Biomaterials. 2009;**30**: 4708

[31] Dai T, Ren W, Wu A. TiO_2 Nanoparticles Applications in Nanobiotechnology and Nanomedicine: Cancer Theranostics of Black TiO_2 Nanoparticles. Wiley; 2020

[32] Thurn KT, Paunesku T, Wu A, Brown EMB, et al. Labeling TiO_2 nanoparticles with dyes for optical fluorescence microscopy and determination of TiO_2-DNA nanoconjugate stability. Small. 2009;**5**: 1318-1332

[33] Wu KCW, Yamauchi Y, Hong CY, Yang YH, Liang YH, Funatsu T, et al. Biocompatible, surface functionalized mesoporous titania nanoparticles for intracellular imaging and anticancer drug delivery. Chemical Communications. 2011;**47**:5232-5253

[34] Bogdan J, Pławińska-Czarnak J, Zarzyńska J. Nanoparticles of titanium and zinc oxides as novel agents in tumor treatment: A review. Nanoscale Research Letters. 2017;**12**:225-240

[35] Wang T, Jiang H, Wan L, Zhao Q, et al. Potential application of functional porous TiO_2 nanoparticles in light-controlled drug release and targeted drug delivery. Acta Biomaterialia. 2015; **13**:354-363

[36] Xu J, Sun Y, Huang J, Chen C, et al. Photo-killing cancer cells using highly cell-specific antibody–TiO2 bioconjugates and electroporation. Bioelectrochemistry. 2007;**71**:217

[37] Ninomiya K, Ogino C, Oshima S, Kuroda S, et al. Targeted sonodynamic therapy using protein-modified TiO_2 nanoparticles. Ultrasonics Sonochemistry. 2012;**19**:607-614

[38] Rengeng L, Qianyu Z, Yuehong L, Zhongzhong P. Sonodynamic therapy, a treatment developing from photodynamic therapy. Photodiagnosis and Photodynamic Therapy. 2017;**19**: 159-166

[39] Kim S, Im S, Park EY, Lee J. Drug-loaded titanium dioxide nanoparticle coated with tumor targeting polymer as a sonodynamic chemotherapeutic agent for anti-cancer therapy. Nanomedicine: Nanotechnology, Biology and Medicine. 2020;**24**:102110-102132

[40] Yamaguchi S, Kobayashi H, Narita T, Kanehira K. Sonodynamic therapy using water-dispersed TiO_2-polyethylene glycol compound on glioma cells: Comparison of cytotoxic mechanism with photodynamic therapy. Ultrasonics Sonochemistry. 2011;**18**(5): 1197-1224

[41] Yuan ZY, Su BL. Insights into hierarchical meso-microporous structured materials. Journal of Materials Chemistry. 2006;**16**:663-677

[42] Yuan ZY, Su BL. Titanium oxide nanotubes, nanofibers and nanowires. Colloids and Surfaces A: Physicochem Eng. Aspects. 2004;**241**:173-183

[43] Wong CL, Tan YN, Mohamed AR. A review on the formation of titania nanotube photocatalysts by hydrothermal treatment. Journal of Environmental Management. 2011;**92**: 1669-1680

[44] Poudel B, Wang WZ, Dames C, Huang JY, et al. Formation of crystallized titania nanotubes and their transformation into nanowires. Nanotechnology. 2005;**16**:1935-1940

[45] Channe SS. Formation of titania nanowires and nanorods on Ti-6Al-4V alloy using electrochemical anodization.

ECS Journal of Solid State Science and Technology. 2023;**12**:014004-001009

[46] Batool SA, Maqbool MS, Javed MA, Niaz A, et al. A review on the fabrication and characterization of titania nanotubes obtained via electrochemical anodization. Surfaces. 2022;**5**:456-480

[47] Wang X, Li Z, Shi J, Yu J. One-dimensional titanium dioxide nanomaterials: Nanowires, nanorods, and nanobelts. Chemical Reviews. 2014;**114**:9346-9384

[48] Kustiningsih I, Slamet PWW. Synthesis of titania nanotubes and titania nanowires by combination of sonication-hydrothermal treatment and their photocatalytic activity for hydrogen production. International Journal of Technology. 2014;**2**:133-141

[49] Lim JH, Choi J. Titanium oxide nanowires originating from anodically grown nanotubes: The bamboo-splitting model. Small. 2007;**3**(9):1504-1507

[50] Tan AW, Pingguan-Murphy B, Ahmad R, Akbar SA. 2012 review of titania nanotubes: Fabrication and cellular response. Ceramics International. 2012;**38**:4421-4435

[51] Lee K, Mazare A, Schmuki P. One-dimensional titanium dioxide nanomaterials: Nanotubes. Chemical Reviews. 2014;**114**:9385-9454

Chapter 2

Review of Titanium Dioxide Preparation and Application

Ali A. Hassan and Alaa H. Taha

Abstract

Advanced oxidation processes (AOPs) have been extensively studied, with the goal of eliminating a wide range of organic pollutants (OPs). Examples of AOPs include ozone, the Fenton process, photo-Fenton, photolysis, photocatalysis, and photolysis of hydrogen peroxide (H_2O_2) and ozone (O_3). AOP without ultraviolet (UV) radiation may not be able to completely eradicate a whole class of OPs. AOPs produce more free radicals when coupled with UV radiation, which improves the effectiveness of the OPs. The specific AOPs and their limitations in light of the complexity of photocatalytic oxidation are briefly discussed in this paper.

Keywords: advanced oxidation processes, photo catalytic, catalyst, oxidation, reaction, optimization and application

1. Introduction

Advanced oxidation processes can totally destroy the organic contaminants to carbon dioxide and water. Amongst the many AOPs, heterogeneous photocatalytic squalor has been originated to be an extremely active treatment skill [1]. The activation of a semiconductor particulate material (such as cadmium sulfide (CdS), titanium dioxide (TiO_2), zinc oxide (ZnO), etc.) by radiation at the right wavelength is the foundation of heterogeneous photocatalysis [2]. The semiconductor particle absorbs photons with sufficient energy to facilitate the conduction of an electron (e^-) from its valence band (VB) to the conduction band (CB), a transition known as band gap energy [3]. This process creates holes in the valence band (h^+) that will function as oxidizing sites, thereby achieving this activation [4]. The heterogeneous photocatalytic degradation can be considered as a significant process among the other approaches including AOPs that are used to treat the wastewater remediation due to its capacity for crushing bunches of inorganic and organic contaminants at the surrounding pressure and temperature [5, 6]. Furthermore, knowledge related to the photocatalytic reactions pertains to interactions of photons that might be having a suitable wavelength with a semiconductor element [7].

Also, photocatalytic degradation procedure under optimum conditions (light power, catalyst loading, pH, and oxidants' concentration) might mineralize the organic contaminations to carbon dioxide (CO_2) and water (H_2O) [8, 9].

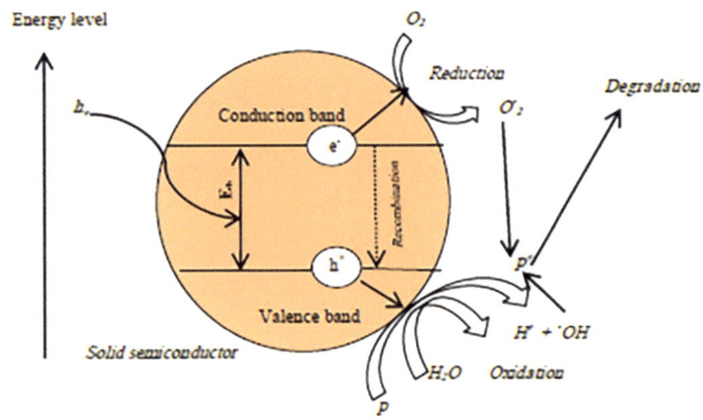

Figure 1.
Standard mechanisms of photocatalysis [10].

A standard mechanism that is related to photocatalysis is provided in **Figure 1**. In the case when the energy of photons (hv) has been comparable to the band gap energy, Eb, related to a photocatalyst, the electrons have been energized and exchanged from the valence band (VB) to the conduction band (CB). Such progression makes openings in the valence band (h) and free electrons (e^-) in the conduction band. Such approach has been stated through the following Equations [11–13]:

Photoexcitation:

$$TiO_2 + hv \rightarrow e^- + h^+ \tag{1}$$

Charge-carrier trapping of e-:

$$e - CB \rightarrow e^-{}_{TR} \tag{2}$$

Charge-carrier trapping of h^+:

$$h^+{}_{VB} \rightarrow h^+{}_{TR} \tag{3}$$

Electron-hole recombination:

$$e^-{}_{TR} + h^+{}_{VB} \rightarrow \text{heat} \tag{4}$$

Photoexcited e^- scavenging:

$$(O_2)ads + e^- \rightarrow O^{\bullet -}2 \tag{5}$$

Oxidation of hydroxyl radicals:

$$OH^- + h^+ \rightarrow {}^{\bullet}OH \tag{6}$$

Photodegradation by $^{\cdot}$OH:

$$R - H + {}^{\bullet}OH \rightarrow R^{'\bullet} + H_2O \tag{7}$$

The e^-_{TR} and h^+_{VB} values given in Eq. (4) are those of the surface-trapped valence electron and the conduction band hole. Without electron acceptors, the e^-_{TR}/h^+_{VB} recombination is extremely anticipated, thus; attendance of the electron scavengers is extremely significant for avoiding such unwanted reaction [14]. Heterogeneous photo-catalytic treatment has been effectively utilized to completely degrade organic contaminations in refinery wastewater by free radical that produced by Eq. (6), the organic photo-catalytic debasement involves the type regarding specific intermediates, for instance, the carboxylic acids and the aldehydes before producing last items CO_2 and H_2O (Eq. 8) [15, 16]:

$$\text{Organic pollutants} \xrightarrow{\text{photocatayst}/hr} \text{Intermediates} \rightarrow CO_2 + \text{H}_2\text{O} \qquad (8)$$

Figure 2 shows the overall complete photocatalytic reactions involving the five steps in the following way [17]:

1. The mass transmission regarding organic contaminant (A) in bulk phase to semiconductor surface.

2. Adsorption related to an organic contaminant on to a photon-activated semiconductor surface.

3. The photocatalysis reaction of adsorbed phase on to a semiconductor surface.

4. Desorption of that of products from a semiconductor surface.

5. The mass transfer of products from the interface area to the bulk fluid (B).

Stages 1 and 5 speak to the mass exchange and the reaction rates that are making physical strides for exchanging contamination between mass and particle surface. In the event these means happen gradually, the mass transfer procedures are constraining and are going to diminish general rate of photocatalytic reaction and conversely [18, 19].

The creation of electrons in the conduction band and positive holes in the valence band often results from the photocatalyst's surface being exposed to a photon with energy, h, equal to or greater than the band gap energy, Ebg [20].

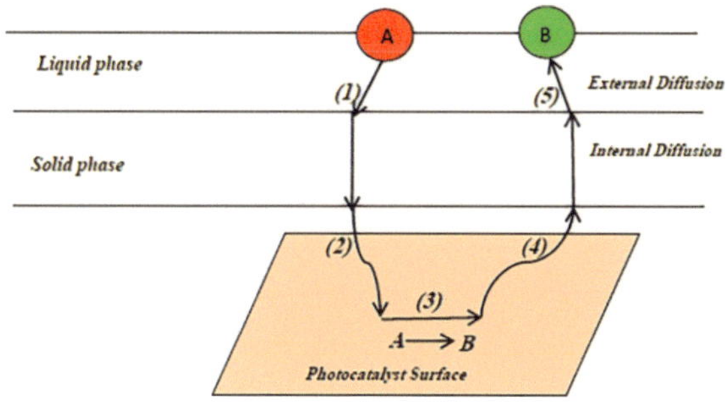

Figure 2.
Steps of heterogeneous catalytic reaction [15].

1.1 Semiconductors' photocatalytic degradation

Tungsten trioxide (WO_3), zinc sulfide (ZnS), tungsten disulfide (WS_2), zirconium dioxide (ZrO_2), cadmium selenide (CdSe), iron (III) oxide (α-Fe_2O_3), cadmium sulfide (CdS), and magnesium sulfide (MgS_2) are just a few of the many metal oxides and chalcogenides that have been used as photocatalysts [21–23]. Regarding the efficiency of semiconductors, the redox capability associated with photogenerated VB opening ought to be positive in order to produce hydroxyl radicals, and it ought to be negative in relation to CB electrons in order to produce superoxide radicals [23].

1.1.1 Titanium dioxide/UV oxidation

TiO_2 has demonstrated great promise as an n-type semiconductor because of its broad band gap (3.2 eV) when exposed to UV light [24]. TiO_2-based materials are important in electrochemistry because of their excellent conductivity and stability in both acidic and alkaline conditions. TiO_2 can be found in three different crystalline phases; the two most prevalent ones are rutile and anatase, and the rutile phase's crystalline size is consistently greater than that of the anatase phase. The third structural type, an orthorhombic structure called brookite, is infrequently used and uninteresting for the majority of applications [25, 26].

Of the three phases, the rutile phase has the highest thermal stability. Above 600°C, the crystalline forms of brookite and anatase undergo a phase transition and become rutile [27, 28]. Whereas the rutile is made up of linear chains with opposite edge-shared octahedral structure, the anatase phase is made up of zigzag chains of octahedral molecules connected to one another [28, 29]. The anatase-to-rutile phase transition normally happens between 600 and 700°C; nevertheless, TiO_2 anatase must be stable at 900°C for some uses [30]. In order to convert the amorphous material into one of the phases of TiO_2, such as rutile, anatase, or brookite, the synthesis methods for TiO_2 normally call for high temperatures. This results in bigger particles and generally nonporous materials [31, 32].

In the process of TiO_2/UV light, titanium peroxide semiconductor will be absorbing the UV light and producing hydroxyl radicals. Especially, throughout UV enlightenment of the TiO_2, the conduction band electrons as well as the valence band gaps have been initially yielded (Eq. (8)). Band electrons interacting with the surface adsorbed subatomic oxygen for yielding superoxide radical anions (Eq. (9)), whereas the band openings are interacting with the water for delivering hydroxyl radical (Eq. (10)) [33]:

$$TiO_2 + hv \rightarrow \mathrm{e}_{cb} + \mathrm{h}^+ \tag{9}$$

$$\mathrm{e}_{cb} + \mathrm{O}_2 \rightarrow \mathrm{O}_2^{-\bullet} \tag{10}$$

$$\mathrm{h}^+ + \mathrm{H_2O} \rightarrow \mathrm{H}^+ + HO^\bullet \tag{11}$$

It is suggested that the first stage of TiO_2-mediated photocatalytic degradation is the production of an (e^-/h^+) pair, which results in the formation of radicals that act as oxidizing agents in photocatalytic oxidation processes: hydroxyl radicals ($^{\cdot}OH$), superoxide radical anions ($O_2^{\cdot-}$), and hydroperoxyl radicals ($^{\cdot}OOH$) [34]. The organic mixtures could experience oxidative degradations through their reactions with the valence bond opening, whereas the hydroxyl and peroxide radicals could experience reductive cleavage via their reactions with electrons. Recently, TiO_2/UV light process is majorly applied with regard to the treatment of wastewater. The major desired conditions regarding such a process feature the lack of mass transfer limitation in the case when the nanoparticles are utilized as photocatalysts, and the operations at the ambient conditions are also using solar irradiations. Furthermore, TiO_2 is costly and promptly provided materials and photogenerated gaps, which are remarkably oxidizing. Furthermore, TiO_2 has been used for oxidation regarding a lot of organic combinations entering into innocuous combinations, for instance, CO_2 and H_2O [35]. The main factors affecting TiO_2/UV light process are the amount that is related to the catalyst, design of the reactor, the initial organic loads, the UV irradiation time, pH solution, temperature, existence of ionic species, and intensity of light. Using extreme measures of the catalyst might reduce energy measure moved into medium due to opacity provided via catalyst particles [36].

TiO_2 is greatly impacted by the pH of the aqueous solution, which also impacts the charge on the particles, the size of the aggregates it forms, and the locations of the valence and conductance bands [37].

Nanocrystalline TiO_2 as a photocatalyst has shown a great many advantages, such as low cost, non-toxicity, high catalytic efficiency, long-term stability, recoverability, etc., and becomes a very promising material for photocatalytic degradation of water pollutants [38].

TiO_2 is known as an excellent photocatalyst due to its properties. It has been used in a variety of applications, for example, electroceramics, glass, and degradation of chemicals in the water and in the air. It has been used in the form of a suspension or in an immobilized water treatment [39].

2. Zinc oxide/UV oxidation

A lot of benefits indicated through ZnO have been specified [40]. That was the focus for most recently the approach that the ZnO has been a semi-conductor with an immediate band gap of 3.44 eV [41, 42]. Magnetic, electrical, and optical properties related to ZnO might be altered or enhanced through using ZnO on a nanoscale [43].

ZnO can be considered as one of the natural friendly materials since it is of high importance to human beings, which lends itself for carrying out daily general applications and which will not be dangerous to the health of humans and the ecological effects [44]. ZnO has a lot of focus on degradation as well as on the complete mineralization of environmental contaminants [45].

It has been suggested that ZnO's photocatalytic capability is equivalent to TiO_2's due to their similar band gap energy (3.2 eV). Moreover, ZnO is less expensive than TiO_2, although TiO_2 has proven to be pricey to use for routine water treatment jobs [46]. The major benefit related to ZnO is the capacity for ingesting sun spectrum ranges and more light quanta in comparison to certain semiconducting metal oxides [47]. The considerable drawbacks of ZnO are photocorrosion and the wide band gap energy. ZnO's light absorption has been constrained in visible light district that arises due to its wide band gap energy. This outcome in the quick recombination regarding the photogenerated charges, therefore, resulted in low photocatalytic productivity [48]. As regards the optimum photocatalytic approach, the organic contaminants have been mineralized into carbon dioxide (CO_2), water (H_2O), and the mineral acids in the attendance of the ZnO particle, also reactive oxidizing types, like oxygen or air [49]. The general approach related to ZnO can be seen in **Figure 3**.

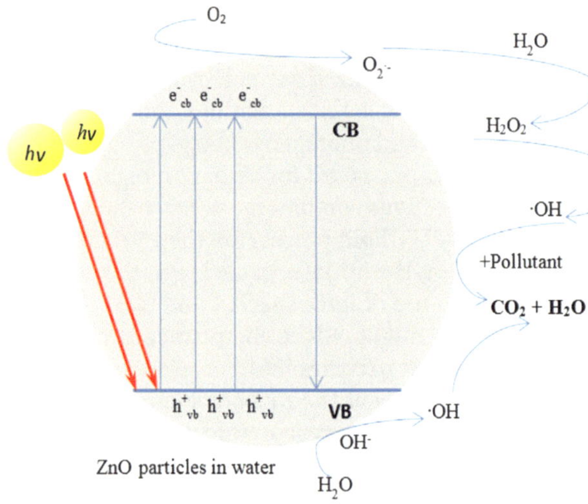

Figure 3.
The general mechanism of photocatalysis [49].

2.1 Important parameters affecting photocatalytic oxidation

The following are the key operating factors that influence the heterogeneous photocatalytic oxidation process's total destruction efficiency [50].

2.1.1 Amount and type of the catalyst

Because of the opacity that the catalyst particles provide, employing too much catalyst lowers the quantity of photoenergy transported in the medium. Therefore, the catalyst concentration should only be utilized up to its optimal value. It should be emphasized that laboratory-scale studies are necessary to determine the optimal value, which will be greatly influenced by the kind and concentration of the pollutant as well as the rate of free radical formation (which is determined by the reactor's operating conditions). The most advantageous catalyst is Degussa P-25 TiO_2. Moreover, catalysts or sensitizers, such as ferrous ions, silver ions, manganese ions, etc., can be employed to raise the photocatalytic oxidation process's treatment effectiveness [51].

2.1.2 Reactor design

Reactor designs typically aim to ensure uniform irradiation of the whole catalyst surface at the intensity of the incident light. This is a significant issue related to the large-scale designs. A high density of the active catalyst in contact with the liquid to be treated inside the reactor, as well as the maximum amount of activated immobilized catalyst exposed to the lighted surface, are essential components of an efficient reactor design [52].

2.1.3 Wavelength of the irradiation

The optimal wavelength for the TiO_2 catalyst, which has a band gap energy of 3.02 eV, is 400 nm. The threshold wavelength for semiconductor catalysts is corresponding to their band gap energy. In certain situations, sunlight can also be utilized to excite the catalyst, which saves a significant amount of money [53].

2.1.4 Initial concentration of the pollutant

Lower starting pollution concentrations are typically preferred. Before the dilution ratio is determined, the drawbacks of treating a high volume of effluent must be examined. When it comes to actual industrial wastes, dilution is frequently required before any degradation is noticeable. Herrmann [17] has said that for actual sewage with a chemical oxygen demand (COD) of 800,000 mg/kg, no degradation was seen and that COD reduction required 1000 times diluting the effluent.

2.1.5 Temperature

Normally, photocatalytic systems are run at normal temperature, but as energy is released during the electron-hole pair recombination process, temperature might rise. Since the pace of reaction drastically decreases after 80°C, interim cooling is advised if the temperature is projected to rise above that point. Exothermic adsorption of the pollutant becomes unfavorable and often acts as the rate-limiting step at temperatures above 80°C, which lowers activity and, consequently, reaction rates [54]. Generally, there is a slight temperature dependence of the degradation rates reported in the 20–80°C range [55].

2.1.6 Radiant flux

The rate of reaction is directly proportional to the radiation intensity; at low intensities, this variation is typically linear; however, beyond a certain intensity magnitude (which depends on the characteristics of the reactor and the effluent), the rate of reaction exhibits a square root dependence on the intensity.

Reduced reliance on radiation intensity is typically ascribed to a higher contribution from the recombination process between electrons and holes produced at high electron densities. The irradiation wavelength is another crucial aspect of the incident light. It is advised to use shorter wavelengths for optimal outcomes. Moreover, as maximum rates are noted at this angle of incidence, the UV light's angle of incidence should always be 90 degrees [56].

2.1.7 Medium pH

The effects of medium pH on photocatalytic oxidation rates are complex, and the observed effects are typically dependent on the kind of pollutant and the semiconductor's zero point charge (zPc), or more precisely, on the electrostatic interaction between the pollutant and catalyst surface. The pollutant's adsorption and, thus, the rates of degradation, will peak close to the catalyst's zPc [57].

Because of an increase in the amount of adsorption in acidic conditions, the rate of photocatalytic oxidation increases at lower pH for particular pollutants, which are weakly acidic [58]. A higher pH may cause some pollutants to hydrolyze in an alkaline environment or to decompose across a specific pH range, which could result in an increase in the rate of photocatalytic oxidation [59].

Fox and Duley [60] have revealed that, throughout the pH range they examined in their investigation, pH had a negligible impact on the degree of deterioration. Since the impact of pH cannot be broadly applied, it is advised that investigations conducted on a laboratory scale be necessary to determine the ideal operating pH values.

2.1.8 Aeration

In order to stop the recombination reaction between the produced positive holes and electrons, the presence of electron acceptors is advised [61]. For this reason, aeration is typically employed since it offers consistent mixing, catalyst suspension in slurry reactors, and an affordable supply of oxygen. Chen and Ray [62] have demonstrated that at partial pressure of 0.2 atm, about 70% of the highest rates of 4-nitrophenol degradation (using pure oxygen) are seen. As a result, air can be utilized safely in place of pure oxygen in commercial operations, significantly lowering operating costs.

2.1.9 Effect of ionic species

Ions can influence the degradation process by UV light absorption, hydroxyl radical ion reactivity, and/or adsorption of pollutants. This is a crucial feature that must be taken into account because, in actual industrial effluents, several salt types will be present at varying concentrations, and these salts are typically ionized.

The chapter contains numerous images that show the effects of different anions and cations. Generally speaking, ions like carbonate (CO_3), bicarbonate (HCO_3), and chloride (Cl) (which greatly influence the adsorption step and partially absorb UV light) can be said to operate as radical scavengers and have an impact on the adsorption process. Ions significantly hinder the degradation process, but only slightly impact the efficiency of degradation when combined with other anions like sulfate, phosphate, and nitrate. Yawalkar et al. have investigated how the ions like sulfate (SO_4), CO_3, Cl, and HCO_3 affect the overall rates of degradation and have found that the negative impacts are seen in the following order: $SO_4 < CO_3 < Cl << HCO_3$ [63].

Author details

Ali A. Hassan* and Alaa H. Taha
Chemical Department, College of Engineering, University of Muthanna, Muthanna, Iraq

*Address all correspondence to: ali.alkhafaji@mu.edu.iq

References

[1] Ahmed IH, Hassan AA, Sultan HK. Study of electro-Fenton oxidation for the removal of oil content in refinery wastewater. IOP Conference Series: Materials Science and Engineering. 2021;**1090**:12005

[2] Alturki SF, Ghareeb AH, Hadi RT, Hassan AA. Evaluation of using photovoltaic cell in the electro-Fenton oxidation for the removal of oil content in refinery wastewater. IOP Conference Series: Materials Science and Engineering. 2021;**1090**:12012

[3] Naeem HT, Hassan AA, Al-khateeb RT. Wastewater-(direct red dye) treatment-using solar Fenton process. Journal of Pharmaceutical Sciences and Research. 2018;**10**:2309-2313

[4] Hassan AA, Naeem HT. Degradation of oily waste water in aqueous phase using solar (ZnO, TiO_2 and Al_2O_3) catalysts. Pakistan Journal of Biotechnology. 2018;**15**(December):927-934

[5] Al-zobai KMM, Hassan AA, Kariem NO. Removal of amoxicillin from polluted water using UV/TiO_2, UV/ZnO/TiO_2, and UV/ZnO. Solid State Technology. 2020;**63**(3):3567-3575

[6] Alakoul KA, Atiyah AS, Azeez MZ, Hassan AA. Photovoltaic cell electro-oxidation for oil removal in oil field produced H_2O. IOP Conference Series: Materials Science and Engineering. 2021;**1090**:12072

[7] Gaya UI, Abdullah AH. Heterogeneous photocatalytic degradation of organic contaminants over titanium dioxide: A review of fundamentals, progress and problems. Journal of Photochemistry and Photobiology C Photochemistry Reviews. 2008;**9**(1):1-12

[8] Diebold U. The surface science of titanium dioxide. Surface Science Reports. 2003;**48**(5-8):53-229

[9] Atiyah AS, Al-Samawi AAA, Hassan AA. Photovoltaic cell electro-Fenton oxidation for treatment oily wastewater. AIP Conference Proceedings. 2020;**2235**:20009

[10] Abeish ABMS. Enhanced photocatalytic degradation of biorefractory pollutants in petroleum refinery wastewater. Curtin University. 2015;**148**(1-2):491-495

[11] Ireland JC, Davila B, Moreno H, Fink SK, Tassos S. Heterogeneous photocatalytic decomposition of polyaromatic hydrocarbons over titanium dioxide. Chemosphere. 1995;**30**(5):965-984

[12] Konstantinou IK, Albanis TA. Photocatalytic transformation of pesticides in aqueous titanium dioxide suspensions using artificial and solar light: Intermediates and degradation pathways. Applied Catalysis B: Environmental. 2003;**42**(4):319-335

[13] Pelizzetti E, Minero C. Mechanism of the photo-oxidative degradation of organic pollutants over TiO_2 particles. Electrochimica Acta. 1993;**38**(1):47-55

[14] Malato S, Blanco J, Vidal A, Richter C. Photocatalysis with solar energy at a pilot-plant scale: An overview. Applied Catalysis B: Environmental. 2002;**37**(1):1-15

[15] Ahmed S, Rasul MG, Brown R, Hashib MA. Influence of parameters on the heterogeneous photocatalytic degradation of pesticides and phenolic contaminants in wastewater: A short review. Journal of Environmental Management. 2011;**92**(3):311-330

[16] Sultan HK, Yousif Aziz H, Hussain Maula B, Hasan AA, Hatem WA. Evaluation of contaminated water treatment on the durability of steel piles. Advances in Civil Engineering. 2020;**2020**

[17] Alamery HRD, Hassan AA. Effect of intensity of light and distance for decolonization in direct red wastewater by photo fenton oxidation. ARPN: Journal of Engineering and Applied Sciences. 2022;**17**(1819-6608):9

[18] Chong MN, Jin B, Chow CWK, Saint C. Recent developments in photocatalytic water treatment technology: A review. Water Research. 2010;**44**(10):2997-3027

[19] Hassan AA, Al-Zobai KMM. Chemical oxidation for oil separation from oilfield produced water under UV irradiation using titanium dioxide as a nano-photocatalyst by batch and continuous techniques. International Journal of Chemical Engineering. 2019;**2019**. DOI: 10.1155/2019/9810728

[20] Ling H, Kim K, Liu Z, Shi J, Zhu X, Huang J. Photocatalytic degradation of phenol in water on as-prepared and surface modified TiO_2 nanoparticles. Catalysis Today. 2015;**258**:96-102

[21] Mills A, Le Hunte S. An overview of semiconductor photocatalysis. Journal of Photochemistry and Photobiology A: Chemistry. 1997;**108**(1):1-35

[22] Fujishima A, Rao TN, Tryk DA. Titanium dioxide photocatalysis. Journal of Photochemistry and Photobiology C: Photochemistry Reviews. 2000;**1**(1):1-21

[23] Al-Hassan AA, Shakir I. Enhanced photocatalytic activity of CuO/NCW via adsorption optimization for refinery wastewater. Iranian Journal of Chemistry and Chemical Engineering. 2024 [Articles in press]

[24] Corma A, Iglesias M, Del Pino C, Sanchez F. New rhodium complexes anchored on modified USY zeolites. A remarkable effect of the support on the enantioselectivity of catalytic hydrogenation of prochiral alkenes. Journal of the Chemical Society, Chemical Communications. 1991;**18**:1253-1255

[25] D'Agata A et al. Enhanced toxicity of 'bulk' titanium dioxide compared to 'fresh' and 'aged' nano-TiO_2 in marine mussels (*Mytilus galloprovincialis*). Nanotoxicology. 2014;**8**(5):549-558

[26] Chen M, Sun X, Qiao Z, Ma Q, Wang C. Anatase-TiO_2 nanocoating of $Li_4Ti_5O_{12}$ nanorod anode for lithium-ion batteries. Journal of Alloys and Compounds. 2014;**601**:38-42

[27] Guo Y, Hu Y, Sigle W, Maier J. Superior electrode performance of nanostructured mesoporous TiO_2 (anatase) through efficient hierarchical mixed conducting networks. Advanced Materials. 2007;**19**(16):2087-2091

[28] Xu J, Li K, Shi W, Li R, Peng T. Rice-like brookite titania as an efficient scattering layer for nanosized anatase titania film-based dye-sensitized solar cells. Journal of Power Sources. 2014;**260**:233-242

[29] Grosso D et al. Highly porous TiO_2 anatase optical thin films with cubic mesostructure stabilized at 700 C. Chemistry of Materials. 2003;**15**(24):4562-4570

[30] Fujimoto M et al. TiO_2 anatase nanolayer on TiN thin film exhibiting high-speed bipolar resistive switching. Applied Physics Letters. 2006;**89**(22):223509

[31] Ramimoghadam D, Bagheri S, Abd Hamid SB. Biotemplated synthesis of

anatase titanium dioxide nanoparticles via lignocellulosic waste material. BioMed Research International. 2014;**2014**

[32] Kominami H et al. Novel synthesis of microcrystalline titanium (IV) oxide having high thermal stability and ultra-high photocatalytic activity: Thermal decomposition of titanium (IV) alkoxide in organic solvents. Catalysis Letters. 1997;**46**(3-4):235-240

[33] Alturki SF, Suwaed MS, Ghareeb A, AlJaberi FY, Hassan AA. Statistical analysis and optimization of mechanical-chemical electro-Fenton for organic contaminant degradation in refinery wastewater. Journal of Engineering Research. 2024

[34] Chen CC, Lu CS, Chung YC, Jan JL. UV light induced photodegradation of malachite green on TiO_2 nanoparticles. Journal of Hazardous Materials. 2007;**141**(3):520-528

[35] Chatterjee D, Dasgupta S. Visible light induced photocatalytic degradation of organic pollutants. Journal of Photochemistry and Photobiology C: Photochemistry Reviews. 2005;**6**(2-3):186-205

[36] Gogate PR, Pandit AB. A review of imperative technologies for wastewater treatment I: Oxidation technologies at ambient conditions. Advances in Environmental Research. 2004;**8**(3-4):501-551

[37] Mills A, Davies RH, Worsley D. Water purification by semiconductor photocatalysis. Chemical Society Reviews. 1993;**22**(6):417-425

[38] Saien J, Delavari H, Solymani AR. Sono-assisted photocatalytic degradation of styrene-acrylic acid copolymer in aqueous media with nano titania particles and kinetic studies. Journal of Hazardous Materials. 2010;**177**(1-3):1031-1038

[39] Ibrahim HA, Hassan AA, Ali AH, Kareem HM. Organic removal from refinery wastewater by using electro catalytic oxidation. In: AIP Conference Proceedings. AIP Publishing; 2023

[40] Nsaif RD, Alturki SF, Suwaed MS, Hassan AA. Lead removal from refinery wastewater by using photovoltaic electro Fenton oxidation. In: AIP Conference Proceedings. AIP Publishing; 2023

[41] Mang A, Reimann K. Band gaps, crystal-field splitting, spin-orbit coupling, and exciton binding energies in ZnO under hydrostatic pressure. Solid State Communications. 1995;**94**(4):251-254

[42] Chen Y et al. Plasma assisted molecular beam epitaxy of ZnO on c-plane sapphire: Growth and characterization. Journal of Applied Physics. 1998;**84**(7):3912-3918

[43] Shrama SK, Saurakhiya N, Barthwal S, Kumar R, Sharma A. Tuning of structural, optical, and magnetic properties of ultrathin and thin ZnO nanowire arrays for nano device applications. Nanoscale Research Letters. 2014;**9**(1):122

[44] Schmidt-Mende L, MacManus-Driscoll JL. ZnO–nanostructures, defects, and devices. Materials Today. 2007;**10**(5):40-48

[45] Anandan S, Ohashi N, Miyauchi M. ZnO-based visible-light photocatalyst: Band-gap engineering and multi-electron reduction by co-catalyst. Applied Catalysis B: Environmental. 2010;**100**(3-4):502-509

[46] Daneshvar N, Salari D, Khataee AR. Photocatalytic degradation of azo

dye acid red 14 in water on ZnO as an alternative catalyst to TiO_2. Journal of Photochemistry and Photobiology A: Chemistry. 2004;**162**(2-3):317-322

[47] Behnajady MA, Modirshahla N, Hamzavi R. Kinetic study on photocatalytic degradation of CI acid yellow 23 by ZnO photocatalyst. Journal of Hazardous Materials. 2006;**133**(1-3):226-232

[48] Gomez-Solís C et al. Rapid synthesis of ZnO nano-corncobs from Nital solution and its application in the photodegradation of methyl orange. Journal of Photochemistry and Photobiology A: Chemistry. 2015;**298**:49-54

[49] Lee KM, Lai CW, Ngai KS, Juan JC. Recent developments of zinc oxide based photocatalyst in water treatment technology: A review. Water Research. 2016;**88**:428-448

[50] Gogate PR, Pandit AB. Sonophotocatalytic reactors for wastewater treatment: A critical review. AICHE Journal. 2004;**50**(5): 1051-1079

[51] Sakthivel S, Neppolian B, Arabindoo B, Palanichamy M, Murugesan V. TiO_2 catalysed photodegradation of leather dye, acid green 16. Journal of Scientific & Industrial Research. 2000;**59**(7):556-562

[52] Mukherjee PS, Ray AK. Major challenges in the design of a large-scale photocatalytic reactor for water treatment. Chemical Engineering Technology. 1999;**22**(3):253-260

[53] Yawalkar AA, Bhatkhande DS, Pangarkar VG, Beenackers AACM. Solar-assisted photochemical and photocatalytic degradation of phenol. Journal of Chemical Technology and Biotechnology. 2001;**76**(4):363-370

[54] Hassan AA, Shakir IK. Kinetic insights into solar-assisted fabrication and photocatalytic performance of CoWO4/NCW heterostructure. Bulletin of Chemical Reaction Engineering & Catalysis. 2024:10

[55] Gouvêa CA, Wypych F, Moraes SG, Durán N, Peralta-Zamora P. Semiconductor-assisted photodegradation of lignin, dye, and kraft effluent by Ag-doped ZnO. Chemosphere. 2000;**40**(4):427-432

[56] Ollis DF, Pelizzetti E, Serpone N. Photocatalyzed destruction of water contaminants. Environmental Science & Technology. 1991;**25**(9):1522-1529

[57] Subramanian MA, Li D, Duan N, Reisner BA, Sleight AW. High dielectric constant in $ACu_3Ti_4O_{12}$ and $ACu_3Ti_3FeO_{12}$ phases. Journal of Solid State Chemistry. 2000;**151**(2):323-325

[58] Oda K et al. p53AIP1, a potential mediator of p53-dependent apoptosis, and its regulation by Ser-46-phosphorylated p53. Cell. 2000;**102**(6):849-862

[59] Choi W. Pure and modified TiO_2 photocatalysts and their environmental applications. Catalysis Surveys from Asia. 2006;**10**(1):16-28

[60] Sakthivel S, Neppolian B, Palanichamy M, Arabindoo B, Murugesan V. Photocatalytic degradation of leather dye over ZnO catalyst supported on alumina and glass surfaces. Water Science and Technology. 2001;**44**(5):211-218

[61] Imahori H et al. Light-harvesting and photocurrent generation by gold electrodes modified with mixed self-assembled monolayers of boron-dipyrrin and ferrocene-porphyrin-fullerene triad. Journal of the American Chemical Society. 2001;**123**(1):100-110

[62] Chen D, Ray AK. Photodegradation kinetics of 4-nitrophenol in TiO_2 suspension. Water Research. 1998;**32**(11):3223-3234

[63] Yawalkar AA, Pangarkar VG, Baron GV. Alkene epoxidation with peroxide in a catalytic membrane reactor: A theoretical study. Journal of Membrane Science. 2001;**182**(1-2):129-137

Chapter 3

Study of Properties and Applications of Titanium Oxide

Ioana Stanciu

Abstract

Titanium dioxide nanocrystals are characterized by the following general considerations: very good chemical stability, transparent thin films, nontoxic, bioactive, low production costs, and the possibility of material regeneration. Some of the general applications of silicon dioxide nanocrystals are: in the manufacture of solar cells, in the decontamination of water, in the decontamination of air, as sensors, gene therapy, and anticorrosion protection. TiO_2 or doped TiO_2 is the most commonly used catalyst in the photocatalytic oxidation of pollutants present in water or air because it is very photosensitive, photostable, biologically and chemically inert, nontoxic, with a good rate of adsorption/desorption of reactants (especially oxygen), being also inexpensive.

Keywords: applications, titanium dioxide, biomaterials, nanomaterials, nanotechnologies

1. Introduction

Biomaterials represent natural, synthetic, or composite materials in contact with living tissues and their biological fluids. They are used to help with the tasks of the affected tissue or the affected functions of a diseased organ.

It is possible to interface the biological environment with medical devices through ID biochips that can teleinteract or improve impaired function of an organ. The premise is that the biomaterial-organism interaction is beneficial [1–3].

Nanomaterial is a fine material made with the help of nanotechnologies.

Nanomaterials are those materials whose component parts have a size between 1 and 100 billionths of a meter (**Figure 1**).

Fortunately, nanomaterials cover a wide range of materials: polymers, metals, and ceramics. They can also have very varied morphologies: spheres, fibers, vanes, dendritic structures, tubes, etc.

Nanotechnology is the set of techniques aimed at the production, manipulation, and use of objects and materials on a nanometric scale (10–9 m), more precisely with dimensions between 1 and 100 nanometers (nm). It can be defined as the ability to transform matter by precisely ordering atom by atom and molecule by molecule to finally produce nanostructures from which nanoproducts will be formed [4–9].

Synthesizing and processing nanostructures have as their object the replacement of various types of organic, inorganic, and biological materials, better than those made until now.

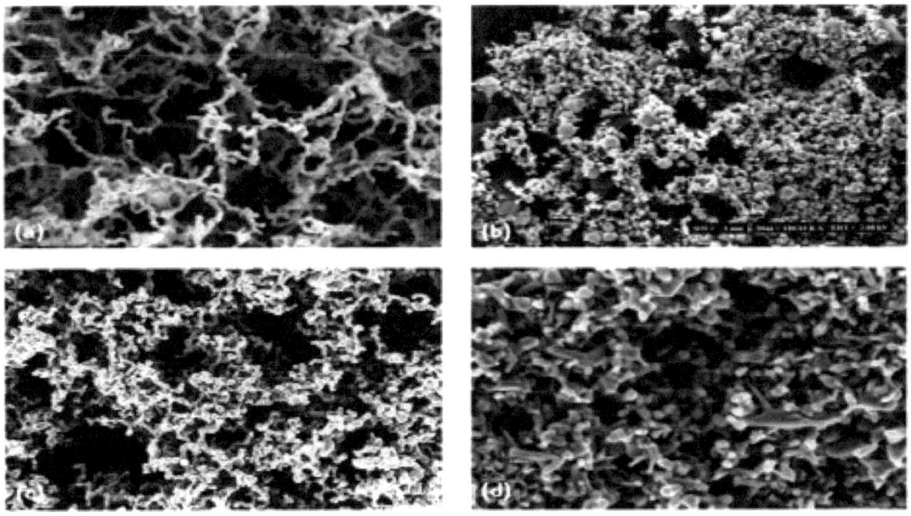

Figure 1.
Appearance of some nanomaterials (nanoparticles-nanopowders): (a) Co; (b) copper oxide; (c) zinc oxide; (d) Ag. https://www.romnet.net/nano/2010.02.03_prezentari/12_Grozescu%20prezentare%20bucuresti.pdf.

TiO_2-based nanomaterials have been intensively studied for water decomposition and the production of hydrogen due to the suitable structure of the electronic band having inseeing the redox potential of water. Other photochemical and photophysical applications of dioxide of titanium include the photolysis of water, the decomposition of organic pollutants in the presence of light, specific catalytic reactions and light-induced superhydrophilicity. Another use interesting of TiO_2-based nanomaterials sensitized with dyes or with metal nanoparticles is represented by the construction of photochromic devices.

Examples of nanomaterials currently being studied or used

- Carbon nanotubes or boron nitride.

- Ceramic nanopowders (silicates or titanium oxide): obtained by vaporizing metallic and/or organic precursors in a flame at high temperature. They are used in the treatment of hardening surfaces, in the production of biocompatible materials for bone implants, and in good electrical conducting polymers.

- Nanofibers, especially carbon: they have electrically conductive properties and high mechanical resistance.

- Glass nanosheets: in the field of optical disks. Obtaining a much higher density of stored information four times by exploiting the deposition of cobalt oxide on the surface of the disk.

- DNA nanofilms: These films have filtering properties useful in the field of environmental protection.

- Nanocrystals: artificial diamond crystals or other natural crystals with electrical properties for making microprocessors.

- Nanocomposites: composite materials with high hardness or transparent.

The advantages of nanometric dimensions

- The structuring of matter at the nanometric level is essential for biological systems
- Nanotechnologies will allow the placement of devices inside cells.
- Nanotechnologies will allow the creation of new materials using self-structuring methods, following the model of those in nature.
- The very high surface/volume ratio characteristic of nanostructures makes them ideal for applications in the fieldcomposite materials, chemical reactions, drug release, energy storage.
- Nanostructured ceramic materials are more mechanically resistant and less fragile.
- Nanoscale catalysis will improve the yield of chemical reactions, combustion in particular, while significantly reducing pollution.
- More than half of the new substances with a therapeutic effect are not soluble in water at the micrometric level, probably dissolving at the nanometric level.
- It thus becomes possible to design new medicines in usable form.
- Nanostructures allow the construction of systems with a significantly increased density of components.
- Electrons will need much shorter times to circulate between components.

1.1 Titanium dioxide

Titanium dioxide, chemical formula TiO_2, also called "titanium white," is an artificial pigment with good covering power, used since 1920. It is chemically inert (**Figure 2**) [10–14].

Formula: TiO_2
Density: 4.23 g/cm^3
Molar mass: 79.866 g/mol.
Boiling point: 2972°C
Melting point: 1843°C
IUPAC number: Titanium dioxide, Titanium (IV) oxide
It generally comes from ilmenite, rutile, and anatase

1.2 Ilmenite

- It is a gray-black mineral, an oxide of iron and titanium, with the chemical formula $FeTiO_3$.

Figure 2.
Titanium dioxide. https://www.samaterials.com/content/application-of-titanium-dioxide-in-the-paper-industry.html.

- Composed of 52.65% of TiO_2 and the rest is FeO.
- It is found in metamorphic and magmatic rocks and crystallizes in the rhombohedral system.
- Ilmenite is the most important titanium ore (**Figure 3**).

1.3 Rutile

- It is a mineral from the group of oxides.
- It has a red-brown color with variants up to black.
- It crystallizes in the tetragonal system.
- It has a hardness of 6–6.5 on the Mohs scale (**Figure 4**).

1.4 Athanasius

- It is found in the form of small and sharp crystals.
- It is an ore that crystallizes in a tetragonal system.
- Its color varies depending on contamination with impurities: dark gray, brown, red-brown, and blue.
- Pure Athanasius is colorless, but it is very rare to find naturally (**Figure 5**).

Figure 3.
Ilmenite. https://en.wikipedia.org/wiki/Ilmenite.

Figure 4.
Rutile. https://rockidentifier.com/wiki/Rutile.html.

Figure 5.
Athanasius. https://ro.wikipedia.org/wiki/Biomaterial#Bibliografie.

1.5 TiO_2 nanocrystals

There is a great interest in obtaining nanomaterials in general and those based on semiconductor oxides in particular, due to their diverse applications. Among semiconductor oxides, titanium dioxide (TiO_2) is the most important material, having been studied a lot in recent years and improved in several variants, one of these being doping with various chemical elements (metallic, nonmetallic ions, or other oxides) following:

a. Influencing the processes involved in the dynamics of quantum mechanisms (generation, movement, recombination of charge carriers—stimulating their participation in the creation of the reactive chemical environment).

b. Expanding the spectrum of the activating radiation toward the visible range, making it possible to activate the material more efficiently in sunlight. In recent years, significant progress has been made in obtaining nanocrystals. Many common materials, such as metals, semiconductors, and magnets, can be obtained from nanocrystals based on colloidal physical processes.

2. Characteristics of titanium dioxide

Titanium dioxide is a fine white powder with very good stability to light, heat, oxidation, and pH changes. It does not present toxicity; it is bioactive. Thin films are transparent.

Acquisition costs are low. Another great advantage is the possibility of regeneration it gives to the materials. Titanium dioxide has the highest refractive index, even higher than that of diamond.

2.1 Applications of titanium dioxide

TiO_2 is used in paints, plastics, or paper to obtain maximum whiteness and opacity, with greater covering power with the ability to mask or hide a substrate.

Domain	Effects
Health	• Anticancer effects • Stimulators of the immune system • Stress protectors
Environment protection	• Air purification • Water treatment for drinking water
Car construction industry	• Antifungal • Antibacterial • Self-cleaning
The glass industry	• Self-cleaning windows
Detection equipment	• Sensors and biosensors

Table 1.
Applications of titanium oxide.

Methods of synthesis	Applications
• Precipitation	TiO_2 accounts for 70% of all pigments worldwide and is widely used to impart whiteness and opacity to various products such as paints, plastics, paper, inks, food products, and toothpaste, as well as cosmetics and skincare that help protect against ultraviolet light.
• Sol-gel	Production of nanomaterials starting from alkoxide solutions or colloidal solutions. Ceramic materials, glass, amorphous and nanostructured materials, and complex oxides.
• Microemulsion	It is used in cosmetics because it improves the adhesion of other ingredients to the skin, being approved for use in cosmetic preparations for lips, eyelids, and skin. TiO_2 is mainly used as a pigment because it has a strong whitening effect. It whitens the prepared compositions or can be used to lighten other pigments. It is also used in make-up products.
• Combustion	
• Hydrothermal	It is a method that is widely used for the production of small particles in the ceramic industry.
• Electrochemical	

Table 2.
Methods of synthesis.

It is also used in the food industry in the manufacture of candies, for white-opaque finishes, or as a base for other colors, for coloring some assortments of cheese, creams, and bakery products. **Table 1** shows the main applications of titanium dioxide.

2.2 Synthesis methods

Table 2 shows the main synthesis methods of titanium dioxide.

3. Testing the immunostimulatory and stress-protective effect

Following the treatment of laboratory animals with undoped and doped titanium dioxide nanocrystals concluded the following:

- Among the classes of immunoglobulins tested, the IgG class presented the highest values compared to those found in the control, suspecting the installation of an immune reactivity expressed by antibodies with a high affinity.
- The increased number of platelets in mice treated with doses of 1 mg undoped TiO_2 indicates an inflammatory process, and it may suspect that undoped TiO_2 in the doses used causes a chronic inflammation unfavorable to the body, while TiO_2-Au and TiO_2-Ag cause an increase in the number of lymphocytes with a stimulating role of the system immune.
- The increased values of IgG in the variants of mice treated with TiO_2-Au, Ag, Fe, and Pt in all experiments support the presumption that doped TiO_2 stimulates immune reactivity [14–20].

4. Titanium dioxide: photocatalyst

A semiconductor is a material whose valence band and conduction band are separated by an energy gap or band gap. When a semiconductor molecule absorbs photons with an energy equal to or greater than its band gap, electrons in the valence band can be excited and can pass into the conduction band, thus generating charge carriers [21]. This semiconductor characteristic of different semiconductor particles, such as TiO_2, WO_3, ZnO, CdS, and SnO_2, allows them to be used in photocatalytic studies. Among these semiconductors, TiO_2 or doped TiO_2 is the most commonly used catalyst in the photocatalytic oxidation of pollutants present in water or air because it is highly photosensitized, photostable, biologically and chemically inert, nontoxic, with a good adsorption/desorption rate of reactants (especially oxygen), being also inexpensive [22]. The study of the physico-chemical principles of the semiconductor-liquid interface demonstrated when using TiO_2 for wastewater treatment is particularly suitable for low concentrations of pollutants [23]. Titanium dioxide has three polymorphs: anatase (tetragonal), rutile (tetragonal), and brookite (orthogonal). There is increasing evidence to suggest that anatase is more active than rutile in the oxidative photocatalytic reaction. Rutile is the thermodynamically stable form of TiO_2, in which anatase and brookite transform upon heating above 500°C or 750°C, respectively [24]. There are many methods to obtain TiO_2 nanopowders, such as chemical vapor deposition (DCV) [25], oxidation of titanium tetrachloride [26], sol-gel technique [27], thermal decomposition, or hydrolysis of titanium alkoxides [28]. The most common TiO_2 used in photocatalysis is the commercial product Degussa P25, produced by flame hydrolysis of $TiCl_4$ at temperatures higher than 1200°C, in the presence of hydrogen and oxygen. P25 is a mixture of anatase and rutile in a ratio of 70:30. The individual P25 particle is nonporous with rounded edges. While the particle size is in the nanoscale, the average diameter of the aggregates is about 1 micron [29].

In numerous investigations, an aqueous suspension of catalyst particles was used. The use of TiO_2 in suspension is somewhat efficient due to the large surface area of the catalyst available for reaction and is common in early-stage photocatalytic research work. Some researchers have even developed pilot-scale wastewater treatment systems using TiO_2 in aqueous suspension [29]. However, the use of suspensions requires the separation and recycling of the ultrafine catalyst from the treated solution. This is usually an inconvenient, time-consuming, expensive process that adds to the overall capital and operating expenses of the facility. In addition, the TiO_2 powder easily agglomerates in the aqueous solution, thus losing its activity. Another problem is that the penetration depth of UV light is very small in the nontransparent titanium suspension due to the strong absorption of the catalyst and dissolved pollutants; thus, the illumination area of the catalyst is still limited. To avoid the separation process, the catalyst can be immobilized on a fixed support. When the catalyst is immobilized, a decrease in the surface area available for reaction is inherently observed. In addition, the reaction takes place at the liquid-solid interface, and the overall rate may be limited to the mass transport of the pollutant on the catalyst surface, with the overall removal efficiency also decreasing [30]. The photocatalyst can be deposited in different substrates, such as reactor walls, a support matrix, or a housing containing the light source [31]. The most common method of TiO_2 immobilization is deposition by immersion in TiO_2 suspension, followed by drying and calcination, because it is simple, cheap, and effective. However, to obtain a uniform film, the coating procedure must be repeated several times to consolidate any significant layer [32–35].

5. Conclusions

Nanomaterials have numerous applications in medicine and the environment, and these are found in various forms: spheres, fibers, pallets, dendritic structures, tubes, etc.

Titanium dioxide, chemical formula TiO_2 has density: 4.23 g/cm^3, molar mass: 79.866 g/mol, boiling point: 2972°C, melting point: 1843°C. It is found as ilmenite, rutile and anatase.

Titanium dioxide is a well-known photocatalytic material that possesses the ability to degrade various organic pollutants and destroy bacteria under the influence of UV irradiation. The band gap energy value of the anatase form of TiO_2 (3.2 eV) is not suitable for solar applications, a fact that practically limits the wide application in the visible range. Development photocatalysts that can be excited in visible light ($\nu > 400$ nm) present a special interest and considerable efforts have been made recently regarding their synthesis methods. One of the methods for obtaining these types of photocatalysts is based on doping them with cations or anions.

Author details

Ioana Stanciu
Faculty of Chemistry, Department of Physical Chemistry, University of Bucharest, Bucharest, Romania

*Address all correspondence to: istanciu75@yahoo.com

References

[1] Rațiu C, Lăzău C, Orha C, Sfîrloagă P, Manea F, Burtică G, et al. Synthesis of hybrid zeolitic materials with TiO2 nanocrystals using solid-solid method. Journal of Optoelectronics and Advanced Materials. 2009;**11**(ISS6):838. ISSN: PRINT: 1454-4164

[2] Stanciu I. Rheological Characteristics of Castor oil used as Biodegradable Lubricant. Oriental Journal of Chemistry. 2020;**36**(5):973

[3] Stanciu I. Dependence Viscosity of Temperature and Shear Rate for Vegetable oil used as Biodegradable Lubricant. Oriental Journal of Chemistry. 2020;**36**(3):563

[4] Rațiu C, Orha C, Sfîrloagă P, Lăzău C, Manea F, Păcală A, et al. Enhancement of natural organic matter removal from surface water using TiO2-modified zeolite. Chemical Bulletin (Beijing, China). 2008;**53**(67):171

[5] Lăzău C, Rațiu C, Sfîrloagă P, Ioițescu A, Miron I, Vlăzan P, et al. Synthesis and characterization of zeolite materials functionalized with undoped and Ag-doped TiO2 nanocrystals. Journal of Optoelectronics and Advanced Materials - Symposia. 2009;**1**. ISSN: Print: 2066-057X

[6] Rațiu C, Lăzău C, Sfîrloagă P, Orha C, Sonea D, Novaconi S, et al. Decontaminate effect of the functionalized materials with undoped and doped (Ag) TiO2 nanocrystals. Environmental Engineering and Management Journal. 2009;**8**(2):237. ISSN: 1582-9596

[7] Rațiu C. Teza de doctorat, Universitatea "Politehnica" Timisoara, Seria 4: Inginerie Chimica, nr. 31, Ed. Politehnica. 2009. ISSN: 1842-8223. ISBN: 978-606-554-002-6

[8] Nitu M, Grozescu I, Lăzău C, Mocanu L, Grozescu AM. Instalatie de sinteza a materialelor nanocristaline prin metoda hidrotermala asistata ultraacustic combinata cu incalzirea in camp de microunde. In: Cerere de brevet de inventie OSIM Nr. A/01020. 2008

[9] Nitu M, Grozescu I, Lăzău C, Mocanu L, Grozescu AM. Sistem de etansare a autoclavei pentru producerea materialelor nanocristaline prin metoda hidrotermala in camp de ultrasunete cu sonotroda imersata. In: Cerere de brevet de inventie OSIM Nr. A/01019. 2008

[10] Grozescu I, Lăzău C, St. Novaconi MS. Instalatie de sinteza a materialelor nanocristaline in camp ultrasonic, prin imersarea sonotrodei. In: Cerere de brevet de inventie OSIM Nr. A/00101. 2007

[11] Lăzău C, Rațiu C, Orha C, Grozescu I, Nițu M, Dabici A. Procedeu de obținere a materialelor hibride pe bază de zeoliți naturali şi nanocristale de TiO2, prin metoda hidrotermală solid-solid în câmp de microunde. In: Cerere de brevet de invenție OSIM A/00546 din. 2009

[12] Lăzău C, Sfîrloagă P, Vlăzan P, Novaconi S, Miron I, Rațiu C, et al. Synthesis and characterization of functional TiO2 nanomaterials. Chemical Bulletin (Beijing, China). 2008;**53**(67):273

[13] Lăzău C, Burada F, Siloşi I, Sfîrloagă P, Rațiu C, Orha C, et al. Immune response of animal organism treated with TiO2 nanocrystals. Scientific and Technical Bulletin. 2008;**13**:48. ISSN: 1582-1021

[14] Lăzău C, Sfîrloagă P, Rațiu C, Orha C, Ioitescu A, Miron I, et al. Synthesis of bioactive materials based on undoped/ doped TiO2 and their nanocrystals

with α/β –cyclodextrins. Journal of Optoelectronics and Advanced Materials. 2009;**11**(ISS7):981. ISSN: PRINT: 1454 – 4164

[15] Corneanu G, Crăciun C, Corneanu M, Lăzău C, Grozescu I. A 7-a ediţie a Seminarului naţional de nanoştiinţă şi nanotehnologie, TiO2-Pt nanopartucles effect on the ultrastructural feautures of Allium sativum sagitatum calus' Bucuresti. 2008

[16] Corneanu CG, Corneanu M, Lăzău C. Actiunea TiO2 asupra cromosomilor la eucariote. Buletinul SNBC. 2008;**36**:115. ISSN: 1584-5532

[17] Corneanu GC, Crăciun C, Corneanu M, Lăzău C, Grozescu I, Siloşi I, et al. The TiO2-Pt nanoparticles implication in the immune response and their interaction with the animal cell. Progress in Nanoscience and Nanotechnologies. 2007;**11**

[18] Corneanu G, Corneanu M, Lăzău C, Grozescu I. The TiO2 nanoparticles effect on the mitotic apparatus in plants (*Alium sativum L.* and *Nigella damascena L.*). In: Understanding Living Systems, XX International Congres of Genetics, Berlin, Germany. 2008

[19] Nitu M, Grozescu I, Lăzău C, Mocanu L, Grozescu AM. Sistem de etansare a autoclavei pentru producerea materialelor nanocristaline prin metoda hidrotermala in camp de ultrasunete cu sonotroda imersata. In: Cerere de brevet Nr. A/01019. 2008

[20] Lăzău C. Teza de doctorat, Universitatea "Politehnica" Timisoara, Seria 4: Inginerie Chimica, nr. 32, Ed. Politehnica. 2009. ISSN 1842-8223. ISBN 978-606-554-010-1

[21] Boer KW. Survey of Semiconductor Physics. NewYork: Van Nostrand Reinhold; 1990

[22] Stanciu I. Study of the Rheological Behavior of Olive oil at Shear Rates Between 3.3 and 120s-1. Oriental Journal of Chemistry. 1 Jan 2023;**39**(1):126

[23] Nozik AJ, Memming R. The Journal of Physical Chemistry. 1996;**100**:13061

[24] Szczepankiewicz SH, Colussi AJ, Hoffmann MR. The Journal of Physical Chemistry. B. 2000;**104**:9842

[25] Agllon JA, Figueras A, Garelik S, Spirkova L, Durand J, Cot L. Journal of Materials Science Letters. 1999;**18**:1319

[26] Jang HD, Jeong J. Aerosol Science and Technology. 1997;**23**:553

[27] Haro-Poniakowski E, Rodriguez-Talavera R, de la Cruz Heredia M, CanoCorona O, Arroyo-Murillo R. Journal of Materials Research. 1994;**9**:2102

[28] Gablenz S, Völtzke D, Abicht H-P, Neumann-Zdralek J. Journal of Materials Science Letters. 1998;**17**:537

[29] Chan CK, Porter JF, Li Y-G, Guo W, Chan C-M. Journal of the American Ceramic Society. 1999;**82**:566

[30] Anderson C, Bard AJ. The Journal of Physical Chemistry. 1995;**99**:9882

[31] Wyness P, Klausner JF, Goswami DY, Schanze KS. Journal of Solar Energy Engineering. 1994;**116**:2

[32] Byrne JA, Eggins BR, Brown NMD, McKinney B, Rouse M. Applied Catalysis B: Environment and Energy. 1998;**17**:25

[33] Diamandescu L, Feder M, Vasiliu F, Tanase L, Sobetkii A, Dumitrescu I, et al. Hydrothermal route to (Fe, N) codoped titania photocatalysts with increased visible light activity. Industria Textila. 2017;**68**:303-308

[34] Duminica FD, Maury F, Costinel D. Elaborari de straturi subtiri de TiO2 prin LPCVD si CVI pentru aplicatii fotocatalitice. Revue Roumaine de Chimie. 2006;**57**(1):52-57

[35] Cifuentes G, Cifuentes L, Kammel R, Torrealba J, Campi A. New methods to produce electrocatalytic lead (IV) dioxide coatings on titanium and stainless steel. International Journal of Materials Research. 2021;**89**(5):363-367

Chapter 4

Perspective Chapter: An Overview of Titanium Dioxide, Uses, Applications and DFT Study of the Optoelectronic Properties of TiO_2 Brookite Clusters

Ife Elegbeleye, Edwin Mapasha, Eric Maluta and Regina Maphanga

Abstract

Titanium dioxide (TiO_2) also known as titania belongs to the class of transitions metal oxides. Titanium dioxide has become a metal oxide of fascinating significance in the research sphere due to its numerous environmental and industrial applications. This chapter presents an overview of the physical, crystal, structural and semiconductor properties of TiO_2 while delving into direct and indirect band gaps, fermi levels in semiconductors, density of states and carrier concentration. The environmental, pharmaceutical, deodorization, photovoltaic and water purification applications of TiO_2 were also discussed. Although TiO_2 clusters have become the focus of several computational studies, typical hardware has a higher processing power, giving way for the simulations of cumbersome systems, some cluster sizes used for some studies are relatively small and are not fit to handle specific problems or complex systems significant for photovoltaic applications. First-principle density functional theory calculation using computational software and GPAW that implements electron density represented on real space grids and the projector-augmented wave method were utilized in this study to investigate the optical and electronic characteristics of TiO_2 brookite clusters. The results of computational investigations on the optical and electrical characteristics of different-sized TiO_2 clusters and intricate systems for the purpose of simulating charge transfer mechanisms in hybrid organic-inorganic photovoltaics and photocatalytic obliteration of contaminants were presented in this chapter.

Keywords: titanium dioxide, photocatalysis, pharmaceutical applications, TiO_2 clusters, density functional theory

1. Introduction

Titanium dioxide (TiO_2), also known as titania, is a member of the transition metal oxides family [1]. Due to its application in pigments, demonstrated capacity as a photocatalyst, and ability to support significant reactions, which are beneficial to the environment, like the production of hydrogen through water splitting, contaminated air and water treatment, and semiconductor material in dye-sensitized solar cells (DSSCs), TiO_2 has attracted increasing attention [2, 3–5]. It has also been extensively utilized in the fields of medicine, environmental protection, and renewable energy [6].

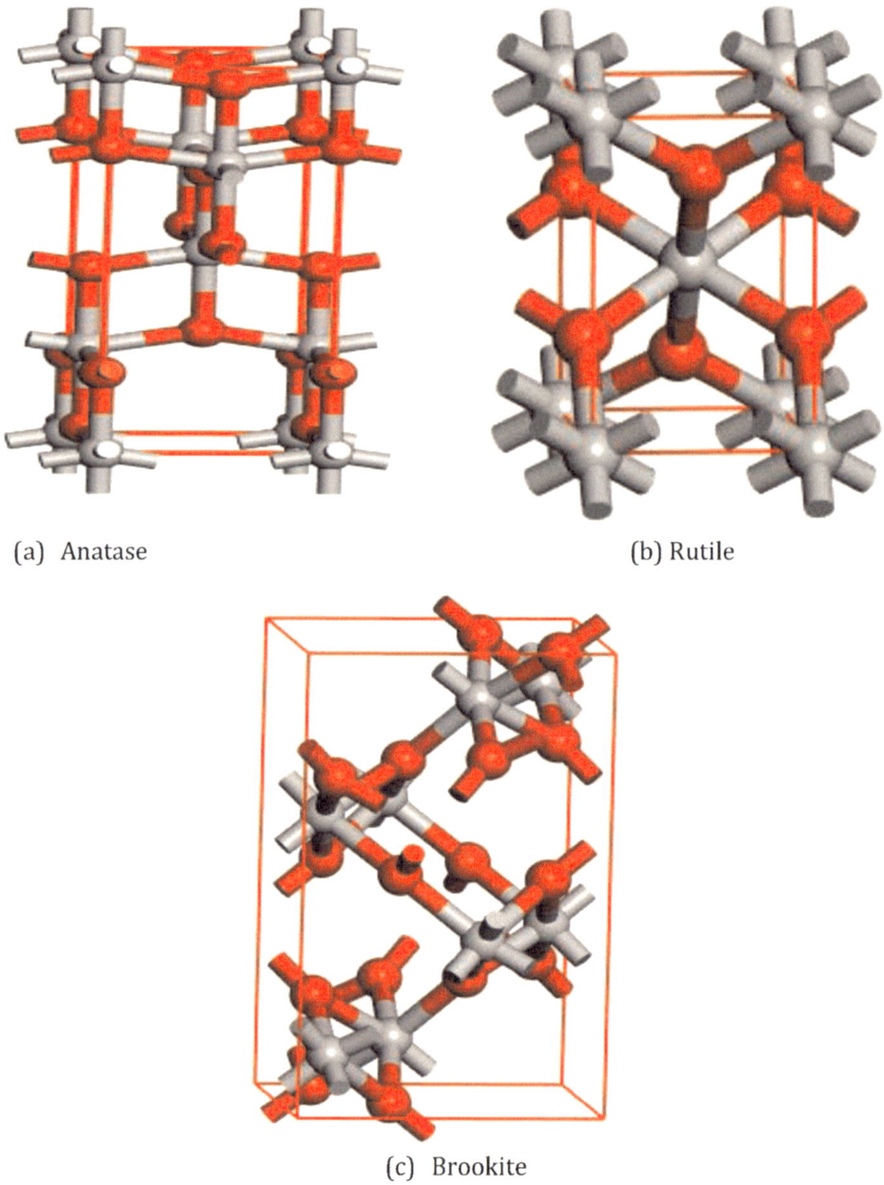

Figure 1.
Crystallographic forms of TiO_2 (a) anatase (b) rutile, and (c) brookite. The color scheme used throughout the study designates the atoms: red balls indicate oxygen atoms and gray balls indicate titanium atoms.

1.1 Polymorphs of TiO_2

Figure 1 shows the well-known polymorphs of TiO_2 structural models. The structures of the three TiO_2 polymorphs were imported from Dassault Systèmes BIOVIA Material Studio Accerlrys Inc. [7]. The structures of anatase, rutile, and brookite presented in **Figure 1** describe titanium atoms bonded by six oxygen atoms in a distorted octahedral configuration. While brookite crystalline structure is basically orthorhombic with each orthorhombic cell comprising eight formula unit. Anatase and rutile building blocks consist of a titanium atom surrounded by six oxygen atoms arranged in an octahedral that is distorted. The octahedral in the rutile phase forms chains that share vertices in the ab-plane and edges along the c-direction, while in the anatase structure, each octahedron shares four edges and forms zigzag chains along the a- and b-directions [1, 3, 8]. The two bonds between the octahedron of the oxygen and titanium atoms are slightly longer in each structure.

2. Applications of titanium dioxide

Titanium dioxide has so many applications ranging from deodorization, air purification, and photocatalysis, as shown in **Figure 2**.

2.1 Environmental improvement applications

Due to the nontoxic nature of TiO_2, it is considered safe for the environment. Additionally, when exposed to sunlight, TiO_2 is a powerful photocatalyst that produces the supra-band gap photon excitation, which improves the breakdown and

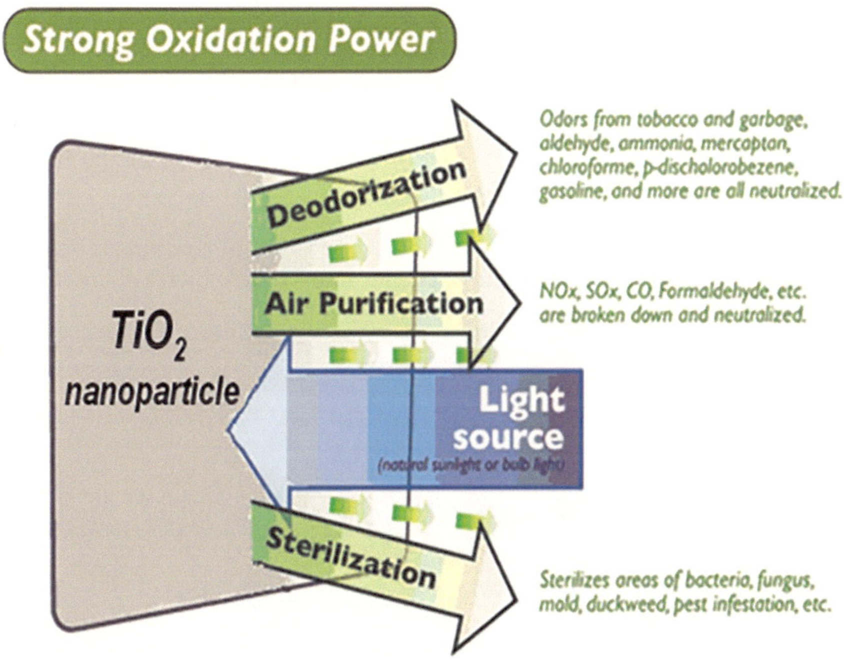

Figure 2.
Applications of titanium dioxide [9].

removal of inorganic matter that is hazardous like SO_2 in the atmosphere and environmental pollutants like NO_2 released into the air by exhaust gas [9].

2.2 Deodorization applications

TiO_2 excellent photocatalytic properties makes it suitable to be utilized in antibacterial and antiseptic compositions, where it breaks down organic pollutants and microorganisms. By breaking down the odor's source, which is ammonia and aldehyde gas (smoke), as shown in **Figure 3**, it essentially targets the source of the stench. When titanium dioxide is exposed to sunlight, it initiates a chemical reaction that breaks down organic poisons, smells, and other substances in the immediate vicinity [9].

2.3 Water purification applications

TiO_2 results in the breakdown of toxins produced by blue green algae and inhibition of other dangerous substances in water, leading to the elimination of organic matter from water, including organic chlorine compounds, trihalomethane, tetrachlorethylene, and methyl-tert-butyl ether trichloroethylene [10]. The following process illustrates how chloroform decomposes.

$$H_2O + \mathrm{CHCl_3} + (1/2)\,O_2 \rightarrow \mathrm{CO_2} + 3\mathrm{HCl} \tag{1}$$

2.4 Pharmaceutical applications

TiO_2 photocatalyst can to break down cell membranes, harden virus proteins, and inhibit virus activation. In the pharmaceutical industry, it is frequently utilized to sanitize equipment that contains viruses. According to a study, TiO_2 can eradicate up to 99.97% of bacteria [11]. TiO_2 can eliminate mildew, suppuration fungus, green suppuration bacillus, and golden grape coccus. The golden grape coccus and coliform have been used to test the sterilizing process. At the start of the experiment, there were 3.2×10^5 golden grape coccus and 3.3×10^5 coliforms, but only 10 remained after the reaction with TiO_2 for 24 hours [11].

2.5 Photocatalytic application

Photons are emitted from the surface of a chemical compound that is light sensitive, when light strikes the compound, a chemical reaction known as photocatalysis

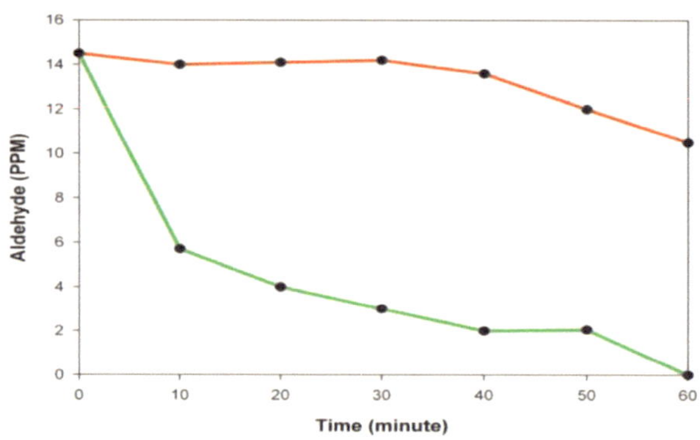

Figure 3.
Decomposition of aldehyde by titanium dioxide [9].

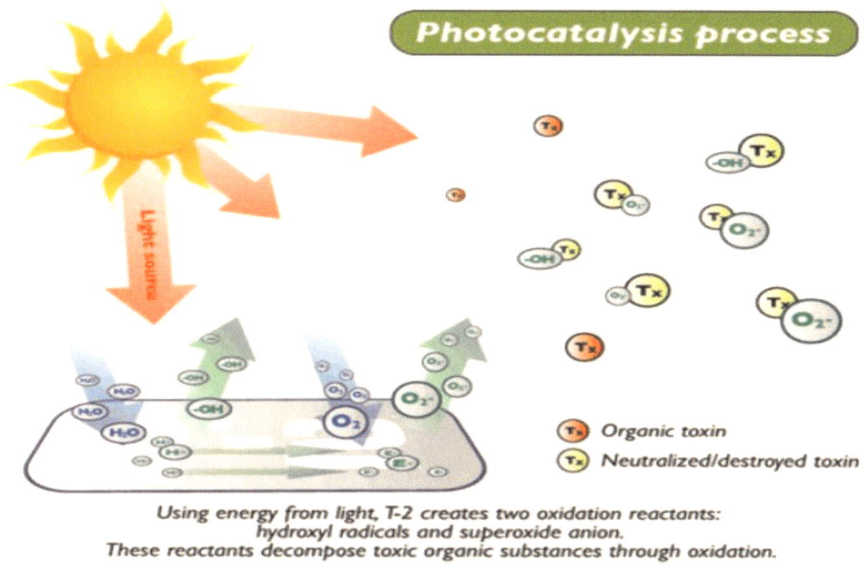

Figure 4.
Photocatalysis of titanium dioxide [9].

occurs. Photons with energy up to or greater than their band gap energy ($\lambda < 385$ nm) are absorbed by TiO_2 photocatalyst when exposed to light. A valence electron will be delocalized and excited to the semiconductor's conduction band as a result. Through one or more types of electron transfer processes, the photoexcited charge carriers can initiate the decomposition of the chemical species absorbed. They can, however, recombine radiatively or non-radiatively and release heat as a byproduct of the input energy, as shown in **Figure 4**. Additives can be added to TiO_2 to increase its efficiency by light illumination, thereby enabling it to become sensitive in the visible region with a decreasing band gap [9].

TiO_2 thin films have been the subject of much research because of their intriguing optical, electrical, and chemical features. TiO_2 nontoxicity and strong stability under light are properties that make it suitable as a dye-sensitized solar cell semiconductor [2, 3, 12, 13]. The most research material for DSSCs photoelectrode application is thin films of TiO_2 despite the existence of metal oxides such as SnO_2 and ZnO semiconductors with large band gaps. This is owing to its less susceptibility to photodegradation when exposed to sunlight. Moreover, the incident photon conversion efficiency (IPCE) and Isc, Voc, and η values are high in DSSCs fabricated using TiO_2 electrodes [14].

3. Properties of TiO_2

3.1 Physical properties of TiO_2

In nature, titanium oxide is found in abundance in mixture with other elements like iron. It can be found in trace levels in rocks that are sedimentary, igneous, and metamorphic. Black hexagonal crystals are the form that titanium dioxide nanoparticles exhibit. TiO_2 is the chemical formula for titanium dioxide, which has a composition of 40.55% oxygen and 59.55% titanium atoms [15]. The electronic configuration is displayed as follows:

Ti → [Ar] $3d^2 4s^2$.

O_2 → [He] $2s^2 2p^4$.

Having a density of 4.23 g/cm^3, a melting point of 1.843°C, and a boiling point of 2.972°C, it has a molar mass of 79.938 g/mol.

3.2 Crystal structural properties of TiO_2

Figure 1 depicted the three well-known titanium dioxides in crystalline phase: rutile, anatase, and brookite. Anatase and brookite are metastable, while rutile is very stable. **Table 1** shows the properties of these polymorphs and as a result of their varying properties, their photocatalytic activities also vary [1].

The space group Pbca describes the unit cell of brookite, which has eight formula units in the orthorhombic cell and an orthorhombic crystalline structure [8, 16, 17]. As seen in **Figure 5**, the formation of brookite can be seen as the connection of three distorted octahedral sharing edges of TiO_6, each with an oxygen atom at the corners and a titanium atom at the center. The crystal's proper chemical composition is provided by the octahedral's extensive sharing of edges and corners. The oxygen atoms are shown in two distinct places in the distorted octahedral [16]. Every titanium and oxygen atom has a distinct bond length [17]. The architecture of DSSCs has made extensive use of rutile and anatase polymorphs [13, 18], while brookite has received less attention and use due to its challenging synthetic process.

3.3 Semiconductor properties of TiO_2

When atoms of large groups come together to form a solid mass or molecules, atomic contact results in changing energy levels because of the mass quantities of

	Lattice parameter (Å)	Space group	System type	Energy band gap (eV)
Rutile	a = 4.594 c = 2.958	$P4_2/mnm\text{–}D^{14}_{4h}$	Tetragonal	3.0
Anatase	a = 3.784 c = 9:515	$I4_1amd\text{–}D^{19}_{4h}$	Tetragonal	3.4
Brookite	a = 9.166 b = 5.436 c = 5.135	$Pbca – D^{15}_{2h}$	Orthorhombic	3.3

Table 1.
Properties of rutile, brookite, and anatase polymorphs of TiO_2.

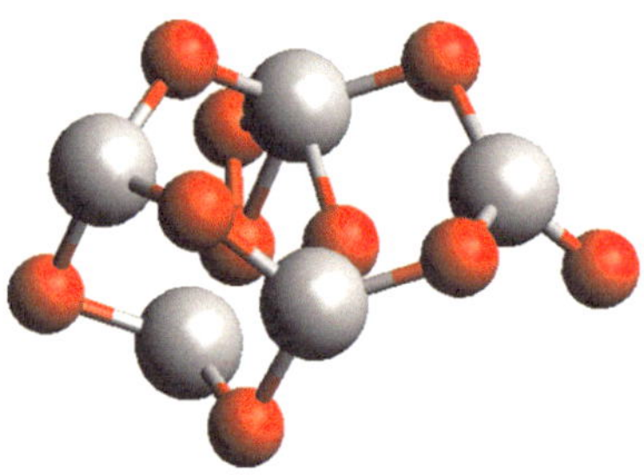

Figure 5.
$(TiO_2)_5$ brookite nanocluster.

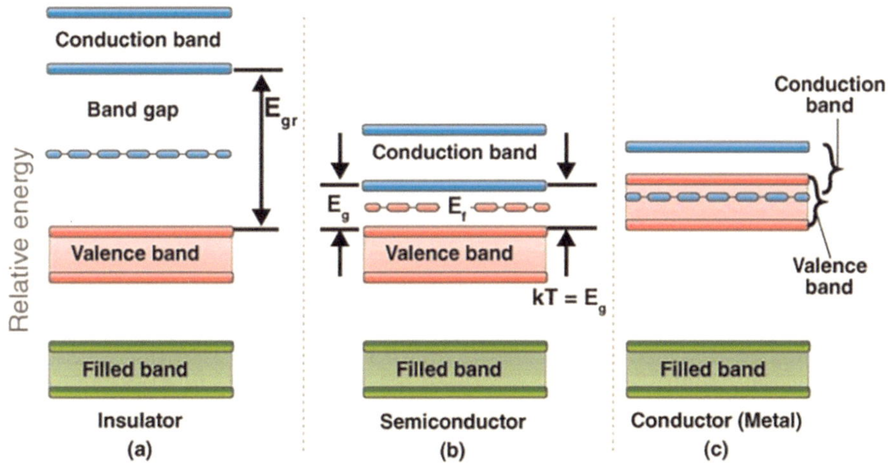

Figure 6.
Valence bands (pink), conduction bands (blue), band gaps, Fermi energy levels for insulators, semiconductors, and conductors Eq. (1).

distinct molecular orbitals, which may be near or even entirely degenerate. Electrons in an isolated atom have definite, defined energy levels. It is claimed that these energy levels form continuous energy bands. The valence band (VB) is the energy band in the highest energy molecular orbital (HOMO) that contains all of the valence electrons. It can be filled entirely or partially. The energy band where mobile charge carriers, either positive or negative, are present is the conduction band (CB). EV stands for the greatest attainable valence-band energy, and EC stands for the lowest attainable conduction-band energy. **Figure 6** shows an illustration of the energy of the band gap, also known as the band gap (EG), which is the difference in energy between the margins of these two bands.

$$E_G = E_C - E_V \tag{2}$$

The conduction band is empty, and the valance band is full at low temperatures. Whereas mobile charge carriers that are negative are electrons, positive mobile charge carriers are holes. The process of conduction in semiconductor materials begins when electrons get sufficient energy to leave a hole in the valence band and jump into the conduction band, where they can flow freely within the crystal lattice [19].

The highest occupied energy level is located at the border between the conduction and the valence in metals, as depicted in **Figure 6**. Very little energy is required by electrons at the top of the valence band to escape into the conduction band [19]. There is a wide energy gap between the conduction bands and valence bands in insulators. It is typically impractical to gain enough energy to excite electrons from the valence band to the conduction band due to the forbidden gap, which is on the order of a few electron volts. Atomically tiny energies are produced in a material via thermal excitations and typical electric circuit voltages. In insulators, the energy difference between the valence and conduction bands cannot be filled with this amount of energy. Semiconductors can be made more conductive by adding tiny amounts of doping material. As a result, the band gap between the valence and conduction bands narrows, leading to a noticeable increase in conductivity. Charge carriers, namely electrons and holes, define semiconductors. The charge carriers show conductivity that is halfway between that of insulators and conductors. They have resistivity in the range of 10^9–10^{-2} Ωcm [20]. As temperature rises, there is an increase in concentration of free electrons and holes. Intrinsic semiconductors, such as silicon and germanium, conduct in chemically pure states.

3.3.1 Direct and indirect band gaps

Phonons and photons interactions (holes and electrons) and energy-momentum (E-k) relationship for carriers in a lattice where momentum and energy must be conserved lead to the concepts of band gaps. Schrödinger equation of an approximate one-electron problem shown in Eq. (4) is solved in order to estimate the energy-momentum (E-k) relationship, or the band structure of a crystalline solid [21].

$$\left(\frac{-h^2}{2m}\nabla^2 + V(R)\right)\varphi(r,k) = E(k)\varphi(r,k) \tag{3}$$

For semiconductor material to be applied in solar cells, lasers, thermoelectric devices, and electronic devices, its band gap is essential. Direct and indirect band gaps are the two types of band gaps seen in semiconductors. In direct-band gap semiconductors, the valance band maximum and conduction band minimum are situated at the same momentum (k) values, as shown in **Figure 7**. There would not be change in momentum values when a hole at the top of the VB recombines with an electron at the bottom of the CB. Radiative transitions are those in which energy is conserved by means of photon emission. As seen in **Figure 3**, the CB minimum and VB maximum in indirect band gap material are located at distinct k-values. Phonons are required for conservation of momentum during the recombination of an electron and a hole in an indirect-band gap semiconductor material [22, 23].

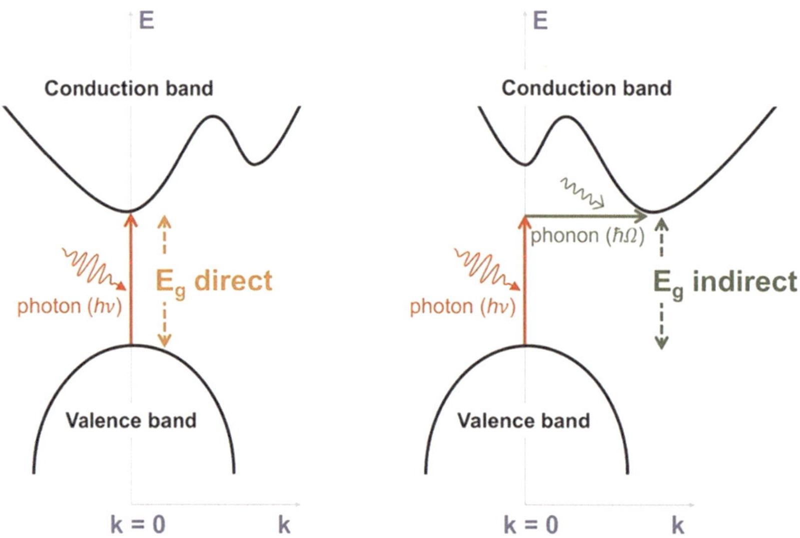

Figure 7.
Photon emission in direct and indirect band gap semiconductors [22].

3.3.2 Fermi levels in semiconductors

A semiconductor material's conduction band is made up of numerous permitted empty energy levels. The likelihood that an electron will fill a level with energy E at thermal equilibrium is represented by the Fermi-Dirac distribution function, or f (E). The Fermi-Dirac distribution function f (E) can be expressed using this (Eq. (4)):

$$f(E) = \frac{1}{1 + exp\left(\frac{E-E_F}{k_B T}\right)} \tag{4}$$

where T is the temperature in Kelvin, E_F is the Fermi energy or Fermi level, and k_B is Boltzmann's constant. Fermi energy is the electrochemical potential of the electrons in a semiconductor material. It stands for the material's average electron energy. The chance that an electron will occupy an energy level E is represented by f(E), but the probability that it will remain empty or that it will have an equivalent hole in the valence band is represented by (1-f(E)) [24].

3.3.3 Density of states and carrier concentration

An energy band is a collection of discrete energy states. Every state in quantum physics corresponds to a unique spin (up and down) and unique solution to the Schrödinger's wave equation for the periodic electric potential function of the semiconductor [22]. The number of electrons (occupied conduction-band levels) for an intrinsic semiconductor is given by Eq. (6), which is the total number of states N(E) multiplied by the occupancy F (E), integrated over the conduction band as presented in (Eq. (5)).

$$n = \int_{E_c}^{\infty} N(E)F(E)dE \tag{5}$$

None or one electron can be present in each state. The density of states, as provided by (Eq. (6)), is the number of states in a narrow range of energy ΔE in the energy bands.

$$D(E) = \frac{numberofstates \in \Delta E}{\Delta E \times Volume} \tag{6}$$

The valence-band VB and conduction-band CB's density of states, denoted by Dc and Dv, respectively, are functions of E, the location of ΔE, according to (Eqs. (7) and (8)).

$$D_c(E) = \frac{8\pi m_n \sqrt{2m_n(E - E_c)}}{h^3} E \geq E_C \tag{7}$$

$$D_v(E) = \frac{8\pi m_p \sqrt{2m_p(E_v - E)}}{h^3} E \leq E_v \tag{8}$$

where mn and mp are electrons and holes effective masses, respectively, averaging over a number of orientations to account for anisotropy. Dc(E) and Dv(E) have dimensions of one number per cubic centimeter per electron Volt. The products Dc (E) dE and Dv(E) dE represent the number of energy levels in the energy range between E and E + dE per cubic centimeter of the semiconductor volume. The separation of charge carriers in semiconductors and the photovoltaic effect are the principles that govern the production of electricity by photovoltaic technology.

TiO_2 performs better as a photocatalyst in nanoparticle form than in bulk structure; extensive research indicates that the charge carrier of a crystalline semiconductor particle behaves quantum mechanically like a simple particle in a box when its diameter is lowered to a threshold radius of 10 nm [25]. Quantized semiconductor particles are more photoactive than microcrystalline semiconductor particles, according to Mill and Le Hunte [26]. This is because the absorption edge blue shifts as particle size decreases, raising the redox potential of the particle's photogenerated holes and electrons.

4. Adsorption of dyes to TiO_2 surfaces and anchor group

One or more anchoring groups anchor dyes to the surface of nanocrystalline semiconductors. The adsorption modes that dyes append to the TiO_2 surface and the electronic coupling that takes place between the dye-excited states and the unoccupied states of the semiconductor are factors that determine the efficiency of DSSCs. There are several methods in which adsorbate oxygen atoms and surface metal atoms can bond an adsorbate to a metal oxide surface. The molecule can bind itself to the surface metal atom via mono (1 M), bi (2 M), or tridentate (3 M) coordination. When there are several metal-oxygen linkages, the number of metal atoms (1 M, 2 M) involved in the adsorption process can also be utilized to differentiate between the adsorption modes.

The most employed anchor groups for attaching sensitizers to semiconductor surfaces are phosphonic acid and carboxylic acid [27, 28]. Some of the potential configurations for the different carboxylic acid adsorption modes are depicted in **Figure 8**.

Recent research on the adsorption of dyes comprising of carboxylic on TiO_2 surfaces indicates that bidentate bridging (BB), shown in **Figure 8c**, is the most favorable adsorption mechanism. In this adsorption mode, one proton is transferred to a neighboring surface oxygen [27, 30–32].

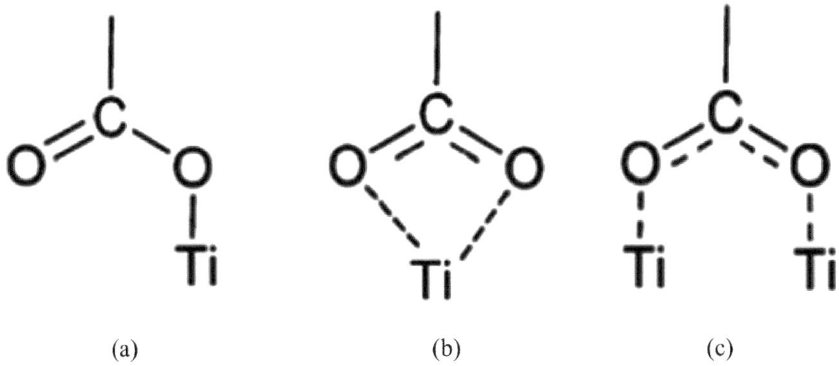

Figure 8.
(a) Monodentate, (b) bidentate chelating, and (c) bidentate bridging [29].

5. Review of related works

Studies using density functional theory on the surfaces of brookite (210) and anatase (101) showed that the building blocks of the two surfaces are comparable. One of the most relaxed and stable surfaces of brookite polymorphs is the brookite (210) surface [33].

A comprehensive density functional theory study on the adsorption geometry of Ru and YE05 sensitizer on TiO_2 substrate for DSSC application was published by Fillipo et al. [30]. The results showed that the dyes functionalized with four carboxylic groups and contained two bipyridine ligands absorbed onto the TiO_2 surface through these carboxylic groups [30].

Density functional theory was used by Prajongtat et al. [34] to examine the electronic and structural characteristics of eight distinct azo dyes (Ar-N=N-Ar, where Ar and Ar′ indicate the aryl group, including benzene and naphthalene skeletons) as well as their absorption into anatase TiO_2. The adsorption energies obtained show that the adsorbed dyes preferentially adopt chelating or monodentate geometries over bidentate bridging configurations. Additionally, the azo compounds containing two carboxyl groups are more selectively coupled to the TiO_2 surface, with the carboxyl group connecting to the benzene moiety rather than the naphthalene moiety [34].

Babara et al. used a vacuum spectrometer and vacuum-tight attenuated total reflection infrared (ATIR-IR) to measure the impact of the anchor and backbone of perylene dye molecules as well as the infiltration of dye molecules unto porous TiO_2 film at a high degree of sensitivity. Their findings showed that dyes with anhydride groups absorb less readily to thin films than dyes with acidic anchor groups. Overall, the simulation findings indicate that the anchoring group has a major effect on the rate of adsorption [35].

With the aid of computer simulations and FT-IR measurements, Chiara et al. examined the energetically favorable TiO_2-adsorption mode of acetic acid as a helpful model for real-world organic dyes. The findings showed that a bridging bidentate adsorption mode was the most durable binding, closely matching the Fourier Transform Infrared Spectroscopy (FT-IR) frequency pattern, for real organic dyes with cyanoacrylic anchoring groups. While the bridging bidentate mode produced a stronger coupling and faster electron injection, the undissociated monodentate adsorption mode for the rhodamine-3-acetic acid anchoring group was shown to be comparably stable. The investigation revealed a relationship between the different electron injection/recombination properties of the oxidized dye and the structural changes caused by the different anchoring groups [36].

Jun et al. [37] investigated alkaline earth metal Ca and N co-doped anatase TiO_2 sheets with exposed (001) facets made using hydrothermal methods. It was confirmed by the X-ray diffractometer and X-ray photoelectron spectroscopy results that the N monodoped TiO_2 is less crystallin than the Ca and N co-doped TiO_2. The study confirms that co-doping Ca and N can successfully lower the generation of recombination centers, increase the efficiency of separating photo-induced electrons and holes, and improve TiO_2's photocatalytic activity based on the hydroxyl radicals (OH) produced during the photocatalytic experiment [37].

In order to alter the photoelectrochemical characteristics of anatase TiO_2, Xu et al. investigated co-doped anatase TiO_2 with transition metals (V or C) and non-metals (N or C). First principles plane wave ultrasoft pseudopotential calculations were used to evaluate the stability and visible light photoactivity, formation energies of the dopant and electronic structures. The results of the investigation demonstrated that co-doping with transition metals makes it easier to increase the p-type dopant concentration. In addition to maintaining the edge of the conduction band oxidation–reduction potential, compensated co-doping lowers the energy gap, increases optical absorbance, improves carrier mobility, and increases conversion efficiency [38].

Puyad et al. studied two model croconate dyes, designated CR1 and CR2, one with an electron-donating substituent (CR1) and the other with an electron-withdrawing group (CR2). They used the periodic density functional theory to study the adsorption of the diketo (-COCO-) groups on the surface of stoichiometric TiO_2 anatase (101). Their findings showed how strongly the acidic group (-COOH) could adhere to the surface of TiO_2. Further theoretical studies also anticipated that the binding strength of the diketone group would be substantial and comparable to that of the -COOH group. This causes a competitive binding of the acid groups on the TiO_2 surface and the diketone groups of croconate dyes [39].

Reports on two croconate dyes, CR1 and CR2, were provided by Chitumalla et al. The researchers used periodic density-functional theory (DFT) simulations and density functional theory to calculate the electrical and optical properties of these dyes. They also examined the adsorption behavior of the two dyes on the TiO_2 (101) anatase surface. Periodic and electronic-structure calculations show that the diketone group of CR1 bonds to the TiO_2 surface more strongly than that of CR2, with a binding strength comparable to that of a typical organic dye. The substituent has a major effect on the croconate dyes' adsorption, optical, and electrical properties [40].

Leonardo et al. [41] reported a periodic density study of a tertiary trimethylamine adsorption on the three most exposed surfaces of stoichiometric anatase TiO_2 nanorods. Following an investigation and characterization of the energetic, structural, and electrical properties, it was found that trimethylamine introduced unique molecular states close to the edge of the TiO_2 valence band [41].

Density functional theory was used by Hao Yang et al. [12] to examine the ruthenium (N3) sensitizer's adsorption behavior on the anatase TiO_2 (001) surface. According to the study's findings, N3 interacts with the (001) surface more strongly than the (101) surface, which causes the (001) surface to have a higher dye coverage. When the N3 sensitizer was adsorbed, its energy gap was reduced, indicating a wider absorption spectra range than when the N3 sensitizer was separated. Furthermore, it was discovered that the TiO_2 (001) surface had a greater conduction band minimum than the TiO_2 (101) surface, indicating a higher open circuit voltage. The findings provided helpful hints and insight into the excellent solar to power DSSCs conversion efficiency including exposed surface TiO_2 (101) nanocrystals [12].

Monique et al. investigated the CO_2 interactions with the (210) surface of brookite TiO_2 by means of first principle simulations on cluster and periodic slab systems. Charge spin density tests were performed to determine the charge transfer to the CO_2 molecule, and the results were compared with the charge transfer to the anatase TiO_2 (101) surface. The study found that the anatase (101) surface and the brookite (210) surface provide CO_2 interactions that are equal in terms of energy. The findings suggested that increasing the amount of oxygen vacancies on the brookite surface might enhance CO_2 absorption. Large levels of CO_2 are verified to be present in the oxygen-deficient brookite based on diffuse reflectance-generated laboratory data and Fourier transform infra-red spectroscopy [33].

Elegbeleye et al. investigated the electronic state energy, optical properties, and energy level alignment of the ruthenium (N3) sensitizer adsorbed on brookite TiO_2 cluster in order to comprehend the electron injection efficiency and kinetics of the dye/TiO_2 complex. The light absorption maximum red shifting to higher wavelength was attributed to the ruthenium N3 dye's absorption on a brookite cluster, which resulted in the distribution and shifting of the lowest unoccupied molecular orbital (LUMO) from the dye to the TiO_2 cluster. The results indicated that the dye-stimulated state of TiO_2 semiconductors can benefit from favorable electron injection. Based on their findings, TiO_2 brookite has the potential to become a new candidate in the DSSC semiconductor market [42].

Many investigations on surface modification of TiO_2 crystals employing atoms or sensitizing dye molecules have been carried out in order to lower the band gap and boost the activity of TiO_2 crystals in the visible and near-infrared parts of the solar spectrum. Anatase and rutile polymorphs of TiO_2 have been used as models in these kinds of studies to improve photocurrent yield and light harvesting in DSSC [12, 24]. The results showed enhanced spectrum responsiveness and increased TiO_2 photocatalytic capabilities. TiO_2 oxide basic research has made substantial use of the surfaces of anatase and rutile polymorphs, which have been employed as a paradigm. The rutile and anatase polymorphs of TiO_2, which are widely used, have been investigated extensively, while the brookite form has gained little or no attention [13].

An investigation on TiO_2 brookite recently suggests that TiO_2 brookite might have higher photocatalytic activity [13]. According to a previous study, brookite's absorption edge is wide and reaches the visible region of the solar spectrum, in contrast to the sharp edges in the visible displayed by the TiO_2 rutile and anatase polymorphs. Considering the dearth of studies on brookite surfaces and their alleged enhanced photocatalytic properties, optimizing photon current density in dye-sensitized solar cells via investigation of dye-brookite TiO_2 interactions is highly desirable.

6. Optical and electronic properties of TiO_2 brookite clusters

6.1 Computational procedures

Using Materials Studio BIOVIA, the bulk structure of the brookite TiO_2 exported from the CASTEP module was optimized to obtain the ground-state structure of the TiO_2 brookite semiconductor [43]. 4x7x7 and 650 eV, respectively, were the convergence energy cut-off and k-points employed for this study. An optimal ground state bulk structure was cleaved to obtain three brookite clusters. The modeled clusters that were cleaved are (i) a brookite nanocluster of size 5 Å in x, y, and z directions; (ii) a cluster with a stoichiometry of $(TiO_2)_n$, where n = 8; and (iii) a cluster with a stoichiometry of (TiO_2) n, where n = 68. Repeating the unit cell supercell by 2 x 2 x 2 Å in the x, y, and z directions led to the formation of $(TiO_2)_{n=68}$ cluster. All the structures were viewed using Avogadro software and exported to GPAW software via the crystallographic information format (cif), for further calculations and analysis.

All DFT calculations were executed within an atomic simulation environment (ASE) using GPAW software [44]. Avogadro was used to display the formations. GPAW is a computer package written in Python that combines the grid space projector-augmented wave (GPAW) with density-functional theory (DFT). **Figures 5–10** show the three TiO_2 brookite clusters that were considered for this study. As mentioned in the preceding section, **Figure 5** shows a $(TiO_2)_5$ brookite nanocluster made up of ten oxygen atoms and five titanium atoms that have been cleaved from the bulk structure of brookite. The structure of brookite $(TiO_2)_8$, which

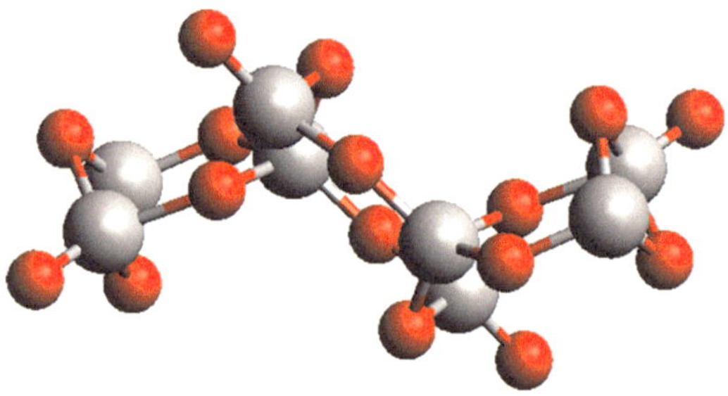

Figure 9.
$(TiO_2)_8$ brookite cluster.

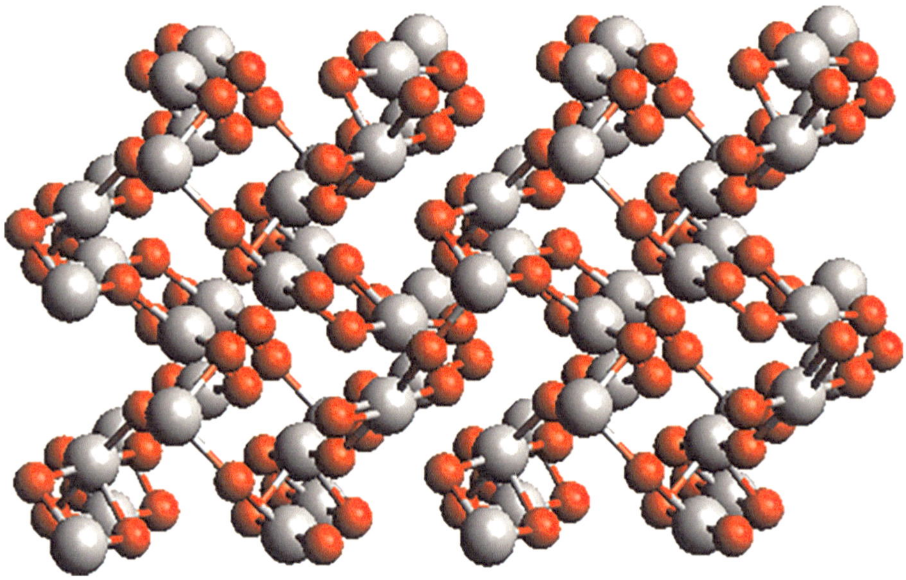

Figure 10.
$(TiO_2)_{68}$ brookite supercell.

is composed of eight titanium and sixteen oxygen atoms, is shown in **Figure 9**. The structure was imported into the Avogadro visualizing interface using the crystallographic mode (cif) without periodicity. A periodic brookite $(TiO_2)_{68}$ supercell measuring 2 x 2 x 2 Å is shown in **Figure 10**. It is made of 136 oxygen atoms and 68 titanium atoms. A cluster or supercell of any size can be formed by repeating unit cells in the x, y, and z directions, which together make up the bulk structure known as the periodic structure.

With the PBE exchange correlation functional, GPAW was used to relax each structure in vacuum. The structures were taken to have converged when all of the atoms were exposed to maximal stresses of roughly 0.05 eV for the non-periodic brookite $(TiO_2)_5$ and $(TiO_2)_8$ cluster models. In the relaxation process, the periodic boundary conditions were applied to the supercell. The atoms in the cluster were rearranged during the relaxing process until the ground state configuration was reached, where the forces converged to a maximum of 0.05 N and the cluster was stable. The nanocluster structures' UV/Vis, total density of states, and partial density of states were computed using the trajectory data obtained from the relaxed structures. The Total Density of States (TDOS) and Partial Density of States (PDOS) from the GPW files were computed using the Perdew–Burke–Ernzerhof (PBE) functional, and the UV/Vis was calculated in vacuum.

6.2 Optical properties of $(TiO_2)_5$ and $(TiO_2)_8$ brookite clusters

The absorption spectra of the $(TiO_2)_5$ and $(TiO_2)_8$ brookite clusters were simulated in vacuum using the TD-DFT method. The TD-DFT calculations were performed while using the PBE exchange correlation functional. The absorption spectra of $(TiO_2)_{68}$ was not computed due to its periodicity. The UV/Vis absorption spectra of $(TiO_2)_5$ brookite nanoclusters and $(TiO_2)_8$ brookite clusters are displayed in **Figures 11** and **12**, respectively. The absorption spectra of $(TiO_2)_5$ seen in **Figure 11** were computed using the hybrid density functional theory B3LYP approximation, and the GPAW was employed to construct the absorption spectra of the complex depicted in **Figure 11** using the PBE exchange correlation functional. The results indicate that both $(TiO_2)_5$ and $(TiO_2)_8$ brookite clusters display absorption in the UV region.

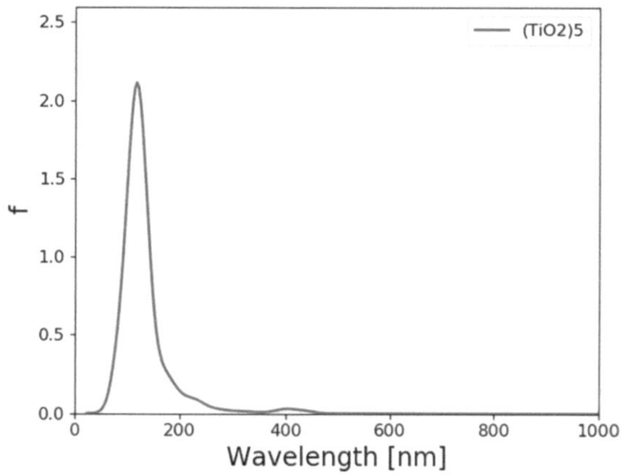

Figure 11.
UV/Vis absorption spectrum for $(TiO_2)_5$ brookite cluster.

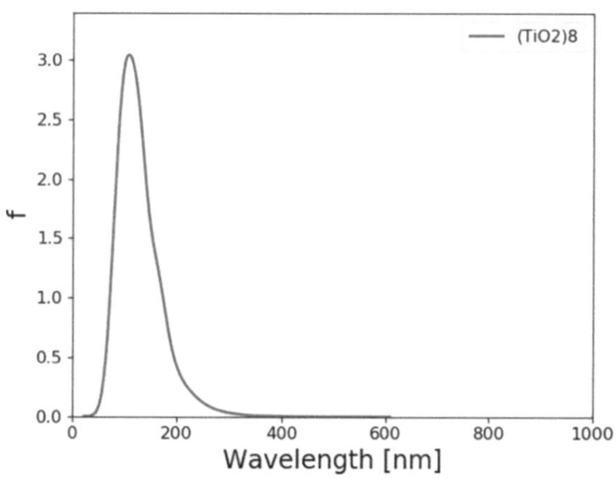

Figure 12.
UV/Vis absorption spectrum for $(TiO_2)_8$ brookite cluster.

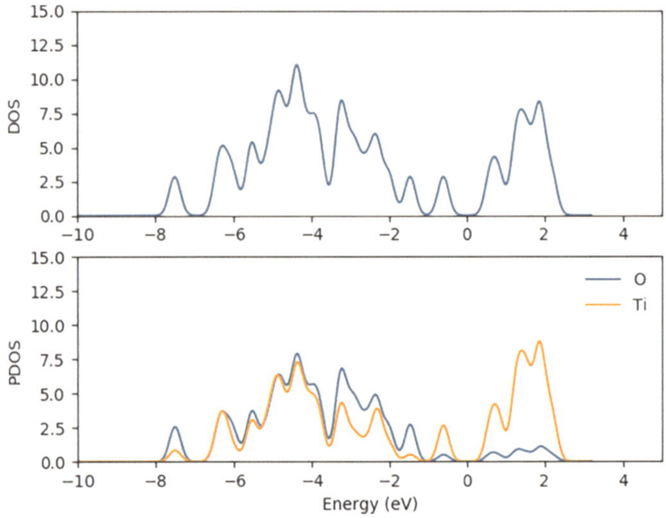

Figure 13.
TDOS (upper) and the estimated DOS (lower) for the $(TiO_2)_5$ nanocluster, where the blue line represents the oxygen contributions to PDOS and the red line represents the titanium atom contributions.

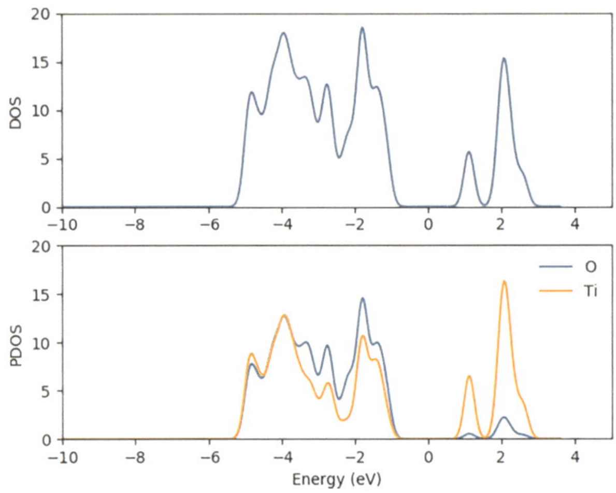

Figure 14.
TDOS (upper) and the estimated DOS (lower) for the $(TiO_2)_8$ nanocluster, where the blue line represents the oxygen contributions to PDOS and the red line represents the titanium atom contributions.

Figure 11 shows more prominent peaks, although this might be because a different functional was used to compute the absorption spectra. $(TiO_2)_5$ and $(TiO_2)_8$ brookite clusters, both show absorption in the UV region of the solar spectrum as shown in **Figures 11** and **12**. Although $(TiO_2)_5$ displays a small absorption peak at 400 nm, notable absorption peaks were mostly situated at 200 nm. An increased peak height observed for $(TiO_2)_8$ brookite absorption spectra relative to the $(TiO_2)_5$ absorption spectra indicates higher absorbance. Due to the wide band gap (3.0–3.2 eV), the absorption spectra of $(TiO_2)_5$ and $(TiO_2)_8$ demonstrate that TiO_2 is primarily sensitive in the UV region of the solar spectrum, which is generally consistent with findings from literature [16].

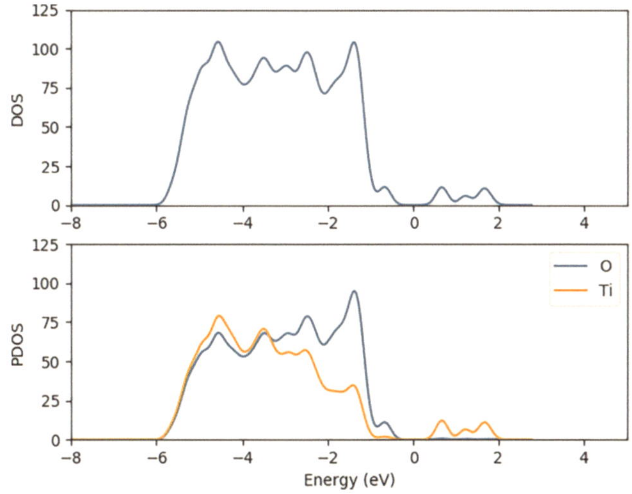

Figure 15.
TDOS (upper) and the estimated DOS (lower) for the $(TiO_2)_{68}$ nanocluster, where the blue line represents the oxygen contributions to PDOS and the red line represents the titanium atom contributions.

6.3 Electronic properties of (TiO_2) n n = 5, 8, 68 brookite clusters

The density of states and projected density of states were computed using GPAW and PBE exchange correction functional in order to better understand the electronic structure of the $(TiO_2)_5$, $(TiO_2)_8$, and $(TiO_2)_{68}$ nanoclusters. The TDOS and PDOS shown in **Figures 13–15** denote the total and partial density of states for $(TiO_2)_5$, $(TiO_2)_8$, and $(TiO_2)_{68}$, respectively. The DOS illustrates the wide band gap that separates the surface valence and conduction bands.

The PDOS signatures for the clusters show that the valence state contributions emanate from the oxygen and titanium atomic orbitals. **Figures 10–12** illustrates that the oxygen 2p atomic orbitals contribute mostly to the highest occupied valence band (VB) state, while the titanium 3d atomic orbitals predominantly contribute to the lowest unoccupied state of the conduction band. With the exception of a tiny amount from the titanium p atomic orbitals, the oxygen p atomic orbitals predominate the valence band. The bulk of the titanium orbitals, especially the d and p ones, contribute to the conduction band; the contributions of the oxygen atoms are minimal.

7. Conclusion

TiO_2 is suited for industrial, environmental, medicinal, deodorization, photovoltaic, and water purification applications due to its intriguing physical, chemical, optical, electronic, and spectrum properties. Rutile and anatase polymorphs of TiO_2 have been thoroughly investigated and used in a variety of applications, particularly for the fabrication of photoanodes for dye-sensitized solar and hybrid organic and inorganic solar cells. As far as we are aware, not much research has been conducted on brookite TiO_2, which has limited its potential for broad use as a semiconductor in solution-processed-based metal organic and hybrid organic–inorganic solar cell device architecture, as well as DSSC photoanodes. Extensive research is required to completely comprehend TiO_2 brookite polymorphs' UV–Vis absorption, electronic excitation energies, and light harvesting efficiency, as well as the related adsorption on dye molecules. Enhancing their optoelectrical and industrial applications will require more knowledge about the optical and electronic properties of the dye/brookite TiO_2 complex. These properties include their formation energies, UV–Vis absorption, HOMO-LUMO energy levels and energy gap, energy level alignment and free energy of electron injection, density of states and projected density of states, photon current densities, and I-V characteristics. Optimizing the optical characteristics of the dye/brookite TiO_2 interface will help to increase the photon to current conversion efficiencies and gain a better knowledge of the absorption mechanisms. This can be accomplished by the thorough study of the TiO_2 brookite polymorph.

Acknowledgements

We express our gratitude to the BwForCluster for Neuroscience, Elementary Particle Physics, and Microsystems Engineering (NEMO), Freiburg, Germany, as well as the Center for High Performance Computing in Cape Town, South Africa, for providing the computational resources needed to implement this study. This chapter was adapted from my dissertation titled "Studies of Interaction of Dye Molecules with TiO_2 Brookite Clusters for Dye-Sensitized Solar Cells Application".

Author details

Ife Elegbeleye[1*], Edwin Mapasha[1], Eric Maluta[2] and Regina Maphanga[3]

1 Department of Physics, University of Pretoria, South Africa

2 Department of Physics, University of Venda, Thohoyandou, South Africa

3 Council for Science and Industrial Research, Pretoria, South Africa

*Address all correspondence to: ifelove778@gmail.com

References

[1] Sungur Ş. Titanium Dioxide Nanoparticles. In: Kharissova OV, Torres-Martínez LM, Kharisov BI, editors. Handbook of Nanomaterials and Nanocomposites for Energy and Environmental Applications. Cham: Springer;; 2021. pp. 3-9. DOI: 10.1007/978-3-030-36268

[2] Nyiko MC, Mpfunzeni R. A review: Simultaneous "one-pot" pollution mitigation and hydrogen production from industrial wastewater using photoelectrocatalysis process. Materials Today Catalysis. 2024;**5**:10052. DOI: 10.1016/j.mtcata.2024.100052

[3] Hayat K, Mansoor UHS. Modification strategies of TiO_2 based photocatalysts for enhanced visible light activity and energy storage ability: A review. Journal of Environmental Chemical Engineering. 2023;**11**:111532. DOI: 10.1016/j.jece.2023.111532

[4] Armakovi'c SJ, Savanovi'c MM, Armakovi'c S. Titanium dioxide as the Most used Photocatalyst for water purification: An overview. Catalysts. 2023;**13**:1-29. DOI: 10.3390/catal13010026

[5] Chauke NM, Mohlala RL, Ngqoloda S, Raphulu MC. Harnessing visible light: Enhancing TiO_2 photocatalysis with photosensitizers for sustainable and efficient environmental solutions. Frontiers in Chemical Engineering. 2024;**6**:1-25. DOI: 10.3389/fceng.2024.1356021

[6] Jafari S, Mahyad B, Hashemzadeh H, Janfaza S, Gholikhani T, Tayebi L. Biomedical applications of TiO_2 nanostructures, recent advances. International Journal of Nanomedicine. 2020;**15**:3447-3470. DOI: 10.2147/IJN.S249441

[7] Material Studio Accerlrys. Dassault Systèmes. BIOVIA. 2016

[8] Diebold U. Structure and properties of TiO_2 surfaces: A brief review. Application Physics. 2003;**76**:681-687. DOI: 10.1007/s00339-002-2004-5

[9] Behzad R, Hamid M. Applications of titanium Dioxdie Nanocoating. Nano-Technology in Environments Conference. 2006:1-3

[10] Ollis DF, Al-Ekabi H. Photocatalytic Purification and Treatment of Water and Air. Elsevier. 1993:511-532

[11] John CI, Petra K, Eugene WR, Robert MC. Inactivation of Escheria coli by titanium dioxide photocatalytic oxidation. Applied and Environmental Microbiology. 1993;**59**:1668-1670

[12] Hao Y, Jia L, Gang Z, Sum WC, Hongda D, Lin G, et al. First principles study of ruthenium (II) sensitizer adsorption on anatase TiO_2 (001) surface. Royal Society of Chemistry Advances. 2015;**5**:60230-60236

[13] Raghvendra SD, Sandesh RJ, Ajinkya BB. Synthesis and characterization of various doped TiO_2 nanocrystals for dye-sensitized solar cells. ACS Omega. 2021;**6**:3470-3482. DOI: 10.1021/acsomega.0c01614

[14] Bandaranayake KMP, Senevirathna MKI, Weligamuwa PMGMP, Tennakone K. Dye sensitized solar cells made from nanocrystalline TiO_2 films coated with outer layers of different oxide materials. Coordination Chemistry Reviews. 2004;**248**:1277-1281. DOI: 10.1016/j.ccr.2004.03.024

[15] DuPont. Titanium Oxide (Titania, TiO_2) Nanoparticles Properties, Applications. AZoNano. 2013. 1-3

[16] Sumaiya IS, Shanawaz A, Allah RA, Sharif Md A. Crystallographic biography on nanocrystalline phase of polymorphs titanium dioxide (TiO_2): A perspective static review. South African Journal of Chemical Engineering. 2024;**50**:51-64. DOI: 10.1016/j.sajce.2024.07.005

[17] Elegbeleye IF, Maluta NE, Maphanga RR. Density functional theory study of optical and electronic properties of $(TiO_2)_{n=5,8,68}$ clusters for application in solar cells. Molecules. 2021;**26**:1-19. DOI: 10.3390/molecules26040955

[18] Di Paola A, Bellardita M, Palmisano L. Brookite, the least known TiO_2 Photocatalyst. Catalysts. 2013;**3**: 36-73

[19] Md AA, Md AS. Semiconductors. International Journal of Advance Multidisciplinary Research. 2020;7:1-8. DOI: 10.22192/ijamr.2020.07.04.001

[20] Kashy ER, Frank NH, Suckling EE, McGrayne SB, Electricity. Encyclopedia Britannica. 2024. Available from: https://www.britannica.com/science/electricity

[21] Sze S, Ng KK. Physics of Semiconductor Devices. Third ed. John Wiley & Sons, Inc; 2007. pp. 1-815. DOI: 10.1002/0470068329

[22] Pedro HMA, Volkringer C, Loiseau T, Tejeda A, Hureau M, Moissette A. Band gap analysis in MOF materials: Distinguishing direct and indirect transitions using UV–vis spectroscopy. Applied Materials Today. 2024;**37**:1-15. DOI: 10.1016/j.apmt.2024.102094

[23] Kelechi I, Chinenye U, Onwumelu AI. The Use of Wide Band Gap Semiconductors in the Production of Electronic Devices and its Feasibility Study in Nigeria – Review. International Journal of Advanced Science and Engineering . 2023;**10**:3410-3421. DOI: 10.29294/IJASE.10.2.2023.3410-3421

[24] Chenming H. Electrons and Holes in Semiconductors. 2009. 1-34

[25] Shikha V. A review on different method of preparation of TiO2 nanoparticles. International Journal of Scientific Research and Review. 2019;**8**: 518-535

[26] Mills A, Hunte SL. An overview of semiconductor photocatalysis. Journal of Photochemistry and Photobiology. 1997; **108**:1-35. DOI: 10.1016/S1010-6030(97)00118-4

[27] Du K, Wang A, Li Y, Xu Y, Li L, Yuan N, et al. The synergistic effect of Phosphonic and carboxyl acid groups for efficient and stable perovskite solar cells. Materials. 2023;**16**:1-11. DOI: 10.3390/ma16237306

[28] Lunlun G, Heng Y, Chengming N, Xuran S, Xiuli W, Mei W. Influence of anchoring groups on the charge transfer and performance of p-Si/TiO_2/Cobaloxime hybrid photocathodes for Photoelectrochemical H_2 production. ACS Applied Material and Interfaces. 2019;**11**:34010-34019. DOI: 10.1021/acsami.9b12182

[29] Adineh M, Pooya T, Mohsen A, Nasser S, Ezeddin M. Fabrication and analysis of dye-sensitized solar cells (DSSCs) using porphyrin dyes with catechol anchoring groups. RSC Advances. 2016;**6**:14512-14521. DOI: 10.1039/c5ra23584g

[30] Fillippo DA, Simona F, Annabella S, Mohammad KN, Michael G. First-principles modelling of the adsorption geometry and electronic structure of Ru

(II) dyes on extended TiO_2 substrates for dye-sensitized solar cell application. Journal of Physical Chemistry. 2010;**114**: 6054-6061. DOI: 10.1021/jp911663k

[31] Nadeem IM, Hargreaves L, Harrison GT, Idriss H, Shluger AL, Thornton G. Carboxylate adsorption on rutile TiO_2(100): Role of coulomb repulsion, relaxation, and steric hindrance. Journal of Physical Chemistry C Nanomaterial Interfaces. 2021;**125**: 13770-13779. DOI: 10.1021/acs.jpcc.1c00892

[32] El-Zohry AM, Agrawal S, De Angelis F, Pastore M, Zietz B. Critical role of protons for emission quenching of Indoline dyes in solution and on semiconductor surfaces. Journal of Physical Chemistry C Nanomaterial Interfaces. 2020;**124**:21346-21356. DOI: 10.1021/acs.jpcc.0c07099

[33] Mengmei P, Niu H, Xingzhang ZJ, Xiaoli Z. Enhanced efficiency of dye sensitized solar cell by high surface area anatase TiO_2 modified P25 paste. Journal of Nanometals. 2013;**2013**:1-6. DOI: 10.1155/2013/760685

[34] Prajongtat P, Suramitr S, Nokbin S, Nakajima K, Misuke K, Hannongbua S. Density functional theory study of adsorption geometries and electronic structures of azo-based molecules on anatase TiO_2 surface for dye-sensitized solar cell applications. Journal of Molecular Graphics and Modelling. 2017; **76**:551-561. DOI: 10.1016/j.jmgm.2017.06.002

[35] Barbara V, Folarin W, Thomas B, Dominic L. Dye bonding to TiO_2: In situ attenuated Total reflection infrared spectroscopy study, simulations and correlation with dye sensitized solar cell characteristics. ACS Publications, American Chemical Society. 2012;**28**:11354-11363. DOI: 10.1021/la302197z

[36] Chiara A, Edoardo M, Mariachiara P, Enrico R, Fillipo De A. Adsorption of organic dyes on TiO_2 surfaces in dye-sensitized solar cells: Interplay of theory and experiment. Physical Chemistry Chemical Physics. 2012;**14**:15963-15974. DOI: 10.1039/C2CP43006A

[37] Jun Z, Lisha Q, Wei F, Junhua X, Zhenguo J. Alkaline- earth metal Ca and N codoped with TiO_2 with exposed (001) facets for enhancing visible light photocatalytic activity. American Ceramic Society. 2014;**97**:2615-2622. DOI: 10.1111/jace.12957

[38] Wenhui X, Xingo M, Tong W, Zhigi H, Wang H, Huang C. First-principles study on the synergistic effects of codoped TiO_2 photocatalysts codoped with N/V or C/Cr. Journal of Semiconductors. 2014;**35**:1-7. DOI: 10.1088/1674-4926/35/10/102002

[39] Puyad AL, Kumar CR, Bhanuprakash K. Adsorption of croconate dyes on TiO_2 anatase (101) surface: A periodic DFT study to understand the binding of diketone groups. Journal of Chemical Sciences. 2012;**124**:301-310. DOI: 10.1007/s12039-012-0229-1

[40] Chitumalla RK, Manho L, Xingfa G, Joonkyung J. Substituent effects on the croconate dyes in dye sensitized solar cell applications: A density functional theory study. Journal of Molecular Model. 2015;**21**:1-8. DOI: 10.1007/s00894-015-2845-4

[41] Leonardo T, Ana B, Munoz-G, Angela A, Michele P. First principles study of trimethylamine adsorption on anatase TiO_2 nanorod surfaces. Theoretical Chemistry Accounts. 2015;**134**:1-11. DOI: 10.1007/s00214-015-1721-8

[42] Elegbeleye IF, Maluta EN, Maphanga RR. Density functional theory

studies of ruthenium dye (N3) adsorbed on a TiO_2 Brookite cluster for application in dye sensitized solar cells. Advances in Quantum Systems in Chemistry, Physics, and Biology, Progress in Theoretical Chemistry and Physics, Springer Nature. 2020;**32**:143-155. DOI: 10.1007/978-3-030-34941-7_8

[43] Clark SJ, Segall MD, Pickard CJ, Hasnip PJ, Probert MJ, Refson K, et al. First principle methods using CASTEP. Zeithchrift fuer Kristallographie. 2005;**220**(5–6):576-570. DOI: 10.1524/zkri.220.5.567.65075

[44] Mortensen JJ, Hansen LB, Jacobsen KW. Real space grid implementation of the projector augumented wave method (GPAW). Physical Review B. 2005;**71**:1-11. DOI: 10.1103/PhysRevB.71.035109

Chapter 5

Perspective Chapter: Electrochromic Efficiency in $Ti_xMe_{(1-x)}O_y$ Type Mixed Metal-Oxide Alloys

Zoltán Lábadi, Noor Taha Ismaeel, Peter Petrik and Miklós Fried

Abstract

Energy-effective smart windows, mirrors, display devices, and automobile sunroofs have been considered as applications of electrochromic materials. This chapter focuses on the electrochromic behavior of Ti-based mixed metal oxides. Transition metal oxides such as Titanium oxide (TiO_2) have been used as promising electrochromic material for this purpose since a smart window contains solid electrolyte and electrochromic material layers (commonly metal oxide layers) sandwiched between transparent conductive layers. However, relatively few publications studied the possible advantages (higher colorization efficiency) of the mixtures of different metal oxides as electrochromic material. This chapter aims to assess the results of investigations of Ti-based multicomponent materials (TiO_2-WO_3, TiO_2-V_2O_5, TiO_2–MoO_3, TiO_2–SnO_2) showing enhanced electrochromic properties compared to the pure TiO_2.

Keywords: mixed metal oxides, reactive sputtering, electrochromic materials, coloration efficiency, TiO_2

1. Introduction

Air conditioning, heating, and ventilation in buildings account for 30–40% of the world's energy consumption [1]. Improving the thermal and optical properties of windows can reduce a building's energy deficit by up to 40% [2]. Therefore, it is important to develop technologies that dynamically control the transparency of windows to reduce energy consumption in buildings. Electrochromic windows are among the favorable solutions to this problem, as they change their light-transmitting properties when exposed to DC bias [3–5]. Dynamic control of sunlight transmission provides an opportunity to reduce energy and lighting costs by 20–50% for commercial buildings [1, 6, 7]. The active components of such electrochromic window technologies are metal oxide layers or organic films [3, 8–11] that display electrochromism, a phenomenon where a material changes its optical properties upon charge injection or extraction. There are many types of such as Titanium Dioxide (TiO_2), Chromium Oxide (CrO), Niobium Pentoxide (Nb_2O_5), Tin Oxide (SnO_2), Nickel Oxide (NiO), Iridium Oxide (IrO_2), Tungsten Trioxide (WO_3), Molybdenum Trioxide (MoO_3), and Vanadium Oxide (V_2O_5) [12–19].

The color change caused by applied direct electric current (DC) is the definition of electrochromic phenomena. Transition metal (Tungsten and Molybdenum) oxide films are the most widely investigated materials for this purpose. The solid-state electrochromic device consists of the electrochromic, charge storage, and electrolyte layers sandwiched between transparent conducting electrodes (TCO).

Electrochromic properties, mainly coloration efficiency, kinetics of the coloration-bleaching process, and cyclic durability of metal oxides strongly depend on its compositional morphological, structural, and characteristics. It is important to study the effect of deposition techniques and growth parameters. Important parameter for EC films is the coloration efficiency (CE) which refers to the optical density change (ΔOD) at a certain wavelength induced when a unit area is injected with charge (Q_d). The CE is calculated using the following equation, and ΔOD is equal with the change of transmittance according to the Beer–Lambert law:

$$\mathrm{CE}(\lambda) = \Delta\mathrm{OD}/\mathrm{Qd} = \log(T_b/T_c/Q_d \tag{1}$$

where T_b is the transmittance of the bleached state, T_c is the transmittance of the colored state, and Q_d is the density of the charge inserted into or extracted from the electrochromic material (cm^2/C, square centimeter per Coulomb). CE has been calculated by specific absorption wavelength (λ) and the transmittances (T_b and T_c) have been dependent on this wavelength. The coloring process evaluated the power requirements by CE and the CE was clear about the electronic efficiency of the ECDs. The result of the CE was presented by a plot of optical change vs. charge density which fitted the linear part of the graph, or alternatively, the relative transmission vs. input charge curves were plotted and CE values can be determined from the fitted exponential curves, see **Figure 1** [20].

Due to possible electron transitions between two sets of electrons, the EC effect can be more pronounced in mixed oxides. Despite this, only a small number of studies can be read about the possible optimization of the EC parameters in mixed oxide-type films. The purpose of this review is to summarize the results of such investigations on Ti-based mixed oxide layers.

2. "Pure" metal oxides

2.1 Tungsten oxide (WO_3)

The most widely studied EC oxide is Tungsten oxide (WO_3), and films of this material have been prepared by several different methods. "Traditional" thin film-making methods include for instance: chemical methods (spin-coating, sol–gel deposition, chemical bath deposition, Langmuir–Blodgett technique, etc.), chemical and physical vapor deposition, electrochemical methods (anodization, plating), see Ref. [10] and references therein. Examples of physical vapor deposition: sputtering [20–22], thermal evaporation [23–27], and pulsed laser deposition [28, 29].

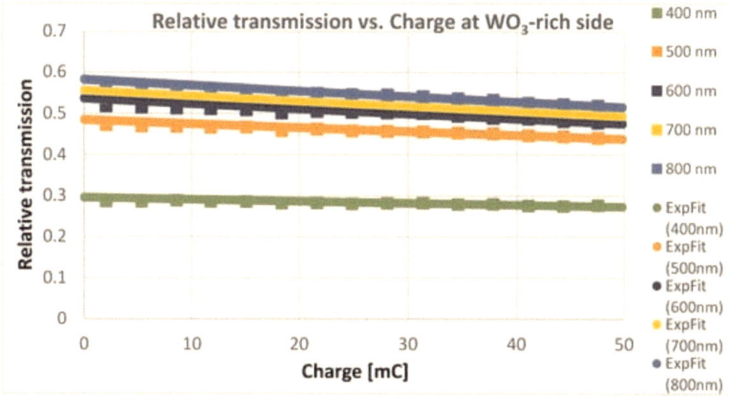

(a)

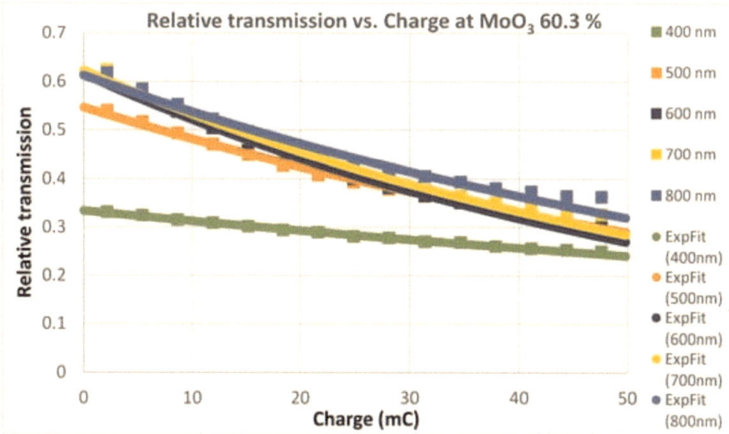

(b)

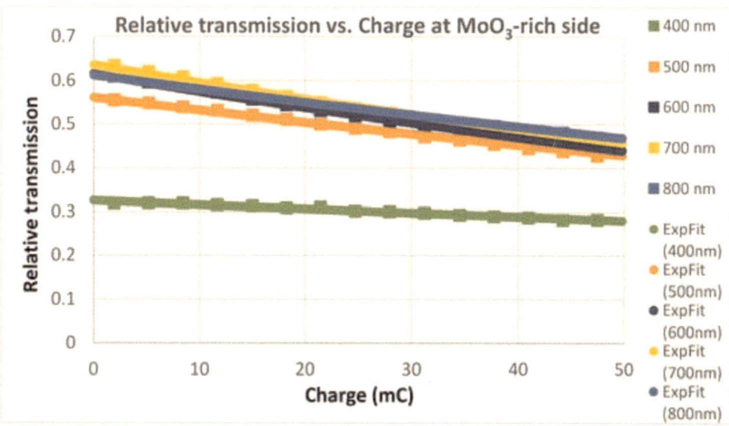

(c)

Figure 1.
Relative transmission vs. input charge curves at five wavelengths for a pure WO3 sample (a), for a WO3-MO3 (Mo-60.3%) mixed sample (b), and for a pure MoO3 sample (c). CE values were determined from the fitted exponential curves (ExpFit). From Ref. [20].

Specific examples of chemical methods are the following: chemical vapor deposition [30–32] and related spray pyrolysis [33–37], and many investigations based on chemical methods to make W oxide films [38–40]. Other researchers used electrodeposition [40, 41], anodization [42–45], and electrophoretic deposition [46]. The CE was found to be more than 60 cm^2/C in the red region in most cases [10, 47].

2.2 Molybdenum oxide (MoO_3)

Electrochromic Mo oxide (MoO_3) shows similar behavior to W oxide, and widespread studies used films prepared by evaporation [48, 49], chemical vapor deposition [50], wet chemical techniques [51, 52], and electrodeposition [53]. CE was measured 34 cm^2/C at 630 nm [49].

2.3 Titanium oxide (TiO_2)

EC properties of Ti oxide (TiO_2) were also extensively studied prepared by the following methods: sputtering [54], chemical vapor deposition [55], and spray pyrolysis [56, 57], various wet chemical techniques [58–60], and anodization [61–63]. The CE was $\sim$25 cm^2/C for reactive DC magnetron sputtering deposited TiO_2 films. TiO_2 films have been also deposited by different chemical techniques and those films were used to determine the CE values, and verified that such films can show the same values of CE [54].

Different deposition methods as radio frequency (RF) reactive sputtering technique [64] and thermionic vacuum arc method have been reported for TiO_2 [65].

3. Mixed oxides

3.1 TiO_2-WO_3

TiO_2 and WO_3 core/shell nanorod arrays were prepared by the combination of hydrothermal and electrodeposition methods, in the work of Cai et al. [66]. The deposition solution was prepared by dissolving Na_2WO_4 salt in deionized water (concentration: 12.5 mM) and adding hydrogen peroxide to the solution maintaining a concentration ratio of 3 with sodium tungstate. Fluorine-doped tin oxide (FTO) glass coated with TiO_2 nanorod array was used as the deposition electrode. Remarkable enhancement of the electrochromic properties was found in the nano-array films. Significant optical modulation (57.2% at 750 nm, 70.3% at 1800 nm, and 38.4% at 10 μm), excellent cycling performance (65.1% after 10,000 cycles) and high CE (67.5 cm^2 C − 1 at 750 nm), fast switching speed (2.4 s and 1.6 s), have been achieved for the core/shell nanorod arrays. Since a larger surface area for charge-transfer reactions was available for ion diffusion, the improved electrochromic properties were mainly attributed to the core/shell structure and the porous space among the nanorod arrays. The presented data made the TiO_2 and WO_3 core/shell nanorod arrays a promising material for practical electrochromic purposes.

Patil et al. [67] have used spray pyrolysis technique at 525°C to deposit TiO_2-doped WO_3 thin films onto FTO coated conducting glass substrates. Tungsten trioxide (WO_3) and titanyl acetylacetonate ($C_{10}H_{14}O_5Ti$) were used as base materials for the deposition of TiO_2-doped WO_3 thin films. The WO_3 powder was dissolved in liquid ammonia at 80°C to obtain an ammonium tungstat solution. The titanyl acetyl acetonate powder ($C_{10}H_{14}O_5Ti$) was mixed with methanol separately at room temperature. The two solutions were stirred in different volume % to form a homogeneous 100 ml precursor solution, at pH = 9. Volume percentage of the dopant varied between 13% and 38% v/v of TiO_2. (For greater than 38% doping percentage of TiO_2 in WO_3 homogeneous solution could not be formed and precipitation dominates.) The thin film samples were uniform, transparent, and strongly adhesive to the substrates. With the help of chronoamperometry (CA), cyclic voltammetry (CV) and chronocoulometry (CC) techniques, electrochromical properties of TiO_2-doped WO_3 thin films have been studied. They concluded that within the studied range, TiO_2 doping enhances the electrochromic performance of WO_3 and samples exhibited increasingly high reversibility with TiO_2 doping concentrations.

Dhandayuthapani et al. [68] reported a low-temperature making of WO_3/TiO_2 films via a combined chemical bath deposition and nebulized spray deposition method. The WO_3 layer influenced the compositional, morphological, structural, and electrochemical properties of TiO_2 films. The layered WO_3 nanoplates on the TiO_2 layer significantly improved the current density of the TiO_2 films. The electrochemical investigation of the annealed WO_3/TiO_2 films displayed a CE of 128.3 cm^2C^{-1}, optical modulation (ΔT) of 78%, and reversibility of 77.2%. Excellent durability for 1000 cycles was exhibited with a fast response of 6 s for coloration and bleaching. This enhancement could be explained by the interconnected nanoplate bundles which accommodate more charges and facilitate faster charge transport. The complementarity of the WO_3–TiO_2 layers leads to efficient electrochromic character [68].

In a recent work, Ashok Reddy et al. [69] investigated the preparation of Titanium dioxide (TiO_2) nanorods/Tungsten oxide (WO_3) hybrid thin films and the effect of nanostructures on the electrochromic characteristic of the films. They deposited WO_3 thin films at the substrate temperature of 400°C. The partial pressures of oxygen were varied. Tungsten nanorods were prepared on FTO coated glass sheets by hydrothermal process. Optimized WO_3 layers were deposited on the TiO_2 nanorod film by sputter deposition. They performed electrochemical, optical, and material analysis on the films using CV, UV–visible spectrometry, x-ray diffraction (XRD), Raman, and x-ray photoelectron spectroscopy (XPS). The enhanced CE of the optimized WO_3 films was attributed to the big active surface area which favored H^+ ions intercalation in the layers. The TiO_2 nanorods/WO_3 hybrid films showed a good electrochemical property in terms of the diffusion coefficient of 1.8×10^{-7} cm^2/s better than those of pure WO_3 (0.6×10^{-7} cm^2/s) and TiO_2 nanorods (0.4×10^{-7} cm^2/s).

Nah et al. [70] demonstrated that homogeneous and well-ordered arrays of $TiO_2 - WO_3$ nanotubes can be layered by anodization of Ti alloys in an ethylene glycol/fluoride-based electrolyte under special electrochemical conditions. They grew nanotube films on different substrates [Ti, Ti-0.2 at% W (Ti-0.2 W), and Ti-9 at% W (Ti-9 W)] by anodization at 120 V in a solution of ethylene glycol with 0.2 Mol HF. The growth time was controlled to achieve a comparable thickness of the layers. Ordered oxide nanotube layers were obtained with a thickness of 1.1–1.2 μm and 85–95 nm tube diameter. These aligned mixed oxide nanotube structures are very good for enhanced electrochromic reactions. It was shown that only small amounts of WO_3 (such as 0.2 at %) can drastically improve the electrochromic properties (cycling stability, contrast, onset potential) of nanotube layer-based devices, see **Figure 2**.

3.2 $Ti_{0.50}V_{0.50}O_x$

Burdis et al. [71] used RF sputtering from metallic targets for deposited thin films of $V_{0.50}Ti_{0.50}O_x$. They used this film as a potential counter electrode in investigating

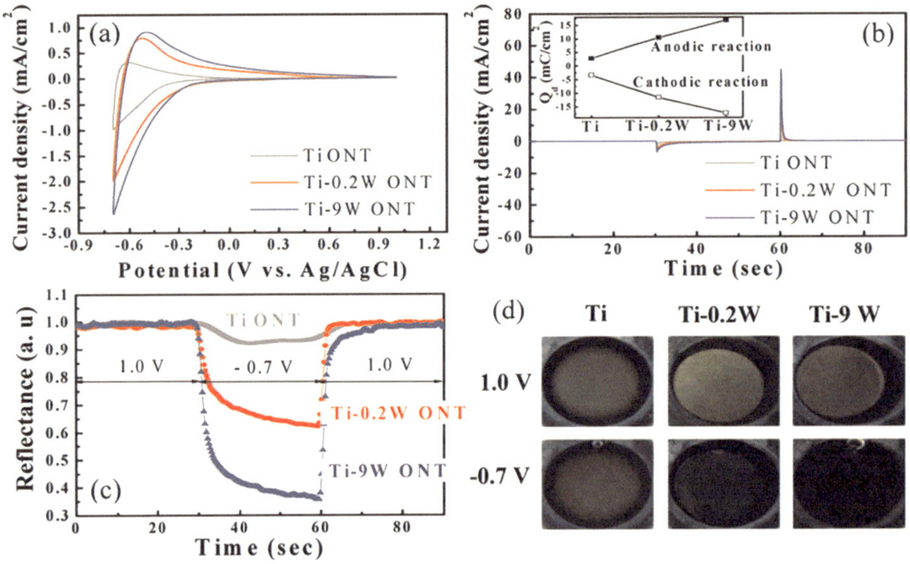

Figure 2.
(a) Cyclic voltammograms of the oxide nanotube (ONT) layers on Ti, Ti-0.2 W, and Ti-9 W performed between −0.7 and 1.0 V with a scan rate of 50 mV in 0.1 Mol $HClO_4$ electrolyte; (b) current density – time curves acquired by pulse potential measurement applied between −0.7 and 1.0 V with 30 s duration; (c) in situ reflectance curves of Ti, Ti-0.2 W, and Ti-9 W ONTs obtained during potential pulsing applied between 1.0 and − 0.7 V; and (d) optical images of the electrochromic effect of the different nanotube surfaces during polarization cycling between 1 and − 0.7 V. The inset of (b) shows integrated charge density (Q_d) for the samples. Reprinted (adapted) with permission from Ref. [70]. Copyright 2008 American Chemical Society.

electrochromic device behavior. They concluded that the film can reversibly store relatively large amounts of charge and it is slightly yellow-looking in transmission, while showing acceptably low electrochromic CE. The electrochemistry of $V_{0.50}Ti_{0.50}O_x$ is found to be simple, in fact rather the same as that of WO_3, for those reasons, they found that this material was considered almost exemplary to use in such a variable transmission device. Charge capacities were measured to 60 mC/cm^2 for 300 nm film thickness. The authors planned to characterize further the electrochromic coloration of these films up to high levels of charge insertion and to determine the effect of repeated charging and discharging on the lifetime of such films.

Marcel et al. [72] used the lamination of two tungsten and vanadium-titanium oxide thin films to reduce the blue absorption of vanadium oxide prepared by roll-to-roll radiofrequency sputtering technique. To produce flexible devices adaptable for eyewear applications, an ITO-coated mylar substrate was used. The working electrode has been set as Tungsten oxide, while the counter-electrodes that were examined vanadium-titanium oxide mixtures. The electrolyte used to assemble both electrodes was a polymer gel lithium ionic conductor constituted of a lithium bis(trifluoromethanesulfonyl)imide (LiTFSI, LiC2F6NO4S2) lithium salt dissolved in propylene carbonate (PC) and incorporated within a photopolymerized acrylate matrix. They investigated the electrochromic properties of the counter electrode at four different atomic ratios of titanium in the (0–100%) range increased at 25% steps. A blue shift effect in the transmittance spectra of as-deposited films was first observed as the titanium amount was increased. In situ, optical behavior was investigated while the potential range for cycling was (1.5-4 V), and the sample with equal proportion of vanadium and titanium displayed a noteworthy neutrality of coloration. Complete devices were prepared with different Ti/V ratios of 0, 1:3, 1:1, and 3:1 in the counter electrode. The film thickness of tungsten oxide has been fixed to 300 nm, and the thickness of vanadium-titanium oxide films has been set based on their respective electrochemical capacity. The obtained electrochromic performances together with their cycling lifetimes and response times were evaluated to find the optimal vanadium-titanium composition.

3.3 TiO_2: MoO_3

Shrestha et al. [73] fabricated self-organized TiO_2–MoO_3 composite oxide nanotubes with tunable characteristics by anodization of a Ti-Mo alloy and these nanotube layers exhibited a considerably enhanced electrochromic color contrast compared with simple TiO_2 nanotubes. They prepared self-organized binary oxide nanotube layers: a single phase Ti-Mo (7 wt%) alloy sheet was polished to a mirror finish, and it was used as a working electrode in a classical anodization assembly. This consists of a traditional 3-electrode system with an Ag/AgCl (3 Mol KCl) reference electrode and a Pt mesh as a counter electrode. The color contrast in terms of reflectivity for the amorphous Ti-Mo-nanotubes is 2.5-fold higher than that of the amorphous TiO2-nanotubes for the same charge density.

Ezhilmaran and Bhat [74] prepared a bilayer electrode with nanoparticulate TiO2 in the bottom layer and randomly oriented MoO3 nanostructures as the top layer to obtain electrochromic device. The TiO_2, MoO_3, and TiO_2/MoO_3 films were prepared using simple solution methods by spin-coating the precursor solutions on conducting FTO substrates. After the spin-coating, the samples were dried and annealed. The heterojunction film was prepared by spin-coating MoO_3 precursor solution on the TiO_2 film. The electrode showed a superior behavior in terms of higher current density and charge storage capacity as well as rate capability as compared to the similar reports in the literature. A color contrast of 38%, switching response of ~2 s and high CE of 72.5 cm2/C were obtained.

To enhance the CE, Ismaeel et al. [75] have performed electrochromic measurements to the full composition range of reactive magnetron sputtered mixed Titanium oxide and Molybdenum oxide (TiO_2-MoO_3). Spectroscopic ellipsometry (SE) has been used to determine and map the composition and optical parameters. To check the results of SE, scanning electron microscopy (SEM) with energy-dispersive x-ray spectroscopy (EDS) has been used. Ti and Mo targets were put separately from each other (see **Figure 3a**), and the indium-tin-oxide (iTO) covered glass sheets and Si-probes on a glass holder (30 cm × 30 cm) were moved under the two separated targets (Ti and Mo) in a reactive argon-oxygen (Ar-O_2) gas mixture, as it can be seen in **Figure 3**. After one sputtering process, all the compositions (from o to 100) % have been achieved by using this combinatorial process, in the same sputtering chamber. Transmission electrochemical cell has been used to determine the CE (the change of light transmission for the unit electric charge) for the mixed metal oxides (TiO_2-MoO_3) that are deposited, see **Figures 3** and **4**. The two maximums in CE (see **Figure 4**) can be explained by the fact that the Ti-rich side was at a much more higher temperature during the deposition process, so the Ti-rich oxide is polycrystalline while the Mo-rich side remains amorphous or nanocrystalline [75]. CE has been considered an important parameter in this study. The maximum value of the CE is 22.2 $cm^2 C^{-1}$ (at λ = 600 nm) at ~60–40% Ti-Mo ratio on the Ti-rich polycrystalline material, while CE is 19.8 $cm^2 C^{-1}$ (at λ = 600 nm) at ~20% - 80% Ti-Mo ratio on the Mo-rich amorphous (or nanocrystalline) material.

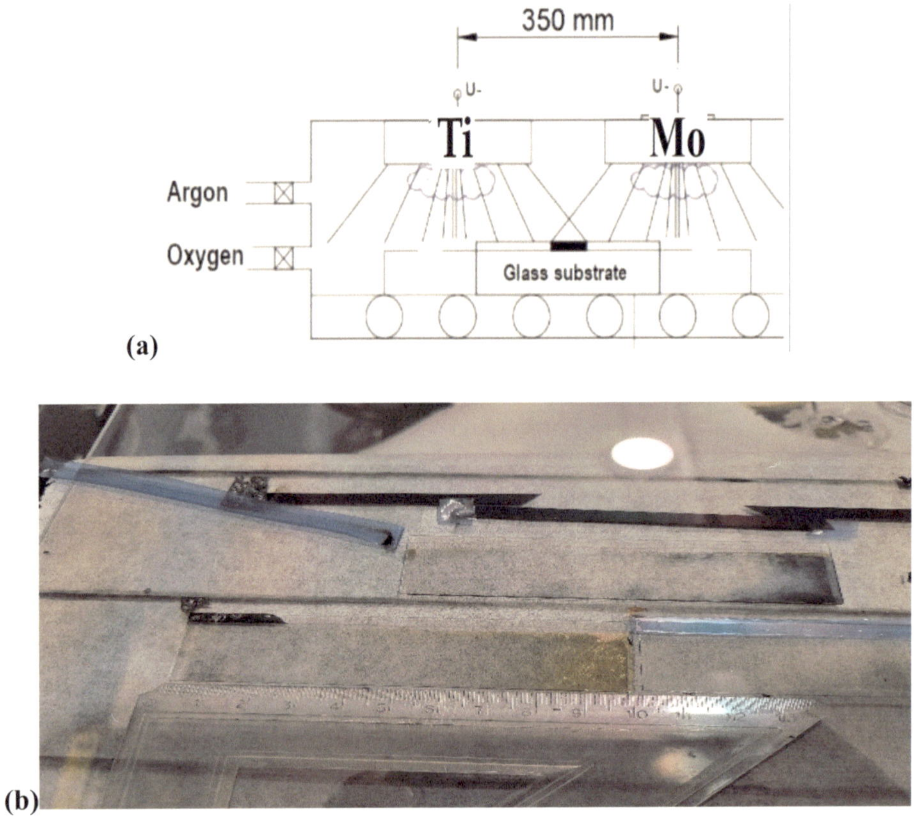

Figure 3.
(a) Schematic of the two targets' arrangements in a closer position (35 cm from each other) and the chamber for the DC magnetron sputtering device; (b) TiO_2-MoO_3 *layers on ITO-covered glasses after-electrochromic-experiments. From Ref. [75].*

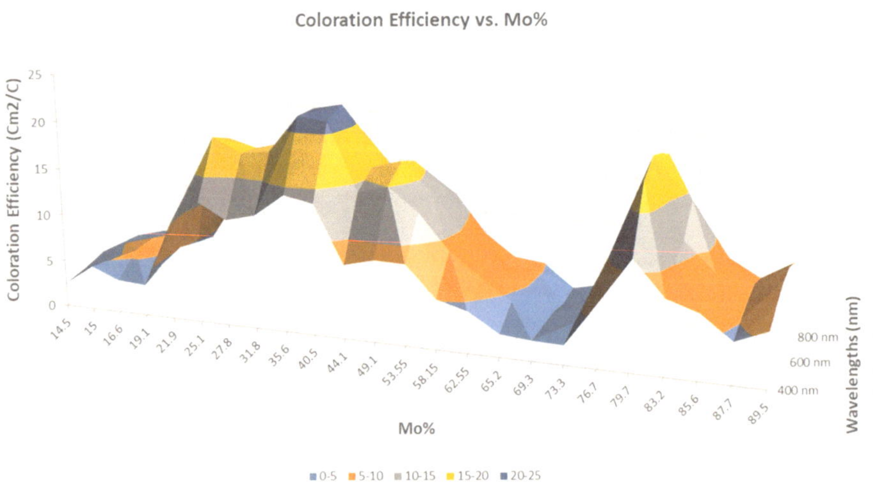

Figure 4.
CE vs. Mo% for wavelengths from (400–800) nm of TiO_2-MoO_3*. From Ref. [75].*

These results are concordant with the results of Habashyani et al. [76] which are published in a recent paper. Using radio frequency magnetron sputtering (RFMS), the authors have grown undoped and Ti-doped vertical nanowall structured MoS_2 thin films and thermally oxidized these films to α-MoO_3 using the following deposition parameters: 45 min oxidation at 380°C under 500 sccm of O_2 gas ambient. Sample names' labeling was the following: undoped MoO_3 - > MBO, Ti:MoO_3 with 20 W RF-power - > MTO_{20}, Ti:MoO_3 with 30 W RF-power - > MTO_{30}, Ti:MoO_3 with 40 W RF-power - > MTO_{40}.

Ti has led to denser nanowall formations. Optical modulation (OM) in the visible region was enhanced with increasing Ti concentration within the coloring potential range of −0.2 to −0.45 V. This is relatively low working voltage, which makes the signals energy efficient in electrochromic materials. The highest Ti-doped MoO_3 (MTO_{40}) and undoped (MTO) samples showed 52.2% and 37.6% OM at λ = 700 nm (47.8% and 25.7%, respectively, at λ = 550 nm) under −0.45 V applied potential. On the other hand, the coloring times for MBO, MTO_{20}, MTO_{30}, and MTO_{40} were 4.7, 4.1, 6.2, and 2.9 s, respectively, while the bleaching durations were 3.1, 1.4, 1.1, and 1.2 s for samples. MBO sample is the undoped while the Mo/Ti ratio in sample MTO_{20} is 79.5, in MTO_{30} is 17.0, in MTO_{40} is 6.7 (highest Ti-doped).

Although the nanowall structure of the highest Ti-doped thin film (MTO_{40}) was destroyed, this thin film showed the best coloring response time and OM at the visible wavelengths. MTO_{20} and MTO_{30} samples, on the other hand, have performed better at longer wavelengths with higher OM and CE. As a result, Ti-doping has a beneficial effect on the electrochromic behavior such as OM, CE, and response times during coloring and bleaching of the MoO_3. These Ti-doped MoO_3 electrochemical characteristics demonstrate the suitability of these materials for device applications.

3.4 TiO_2: SnO_2

Ismaeel et al. [77] determined the optimal composition of reactive magnetron-sputtered combinatorial mixed layers of titanium oxide and tin oxide (TiO_2-SnO_2) for electrochromic purposes. SE was used to obtain the thickness and composition maps of the sample. They also compared the performance of different optical models, such as 2-Tauc–Lorentz multiple oscillator model (2 T–L) or the Bruggeman effective medium approximation (BEMA) to map the sample parameters. To check the results of SE, SEM, with EDS was used. It was shown that in the case of molecular-level mixed layers, 2 T–L is better than an EMA-based optical model. By using SE, the

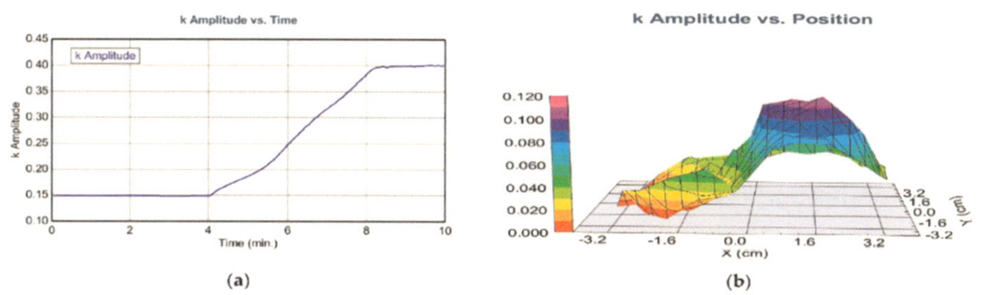

Figure 5.
(a) The imaginary part of the refractive index (k Amplitude) as a function of time for highly conductive Si in the liquid cell during coloration (time-scan, simple 2-layer Cauchy model). From 0 to 4 min, there is low absorption, however, from 4 to 8 min, there is a growing absorption; and (b) Map of the k parameter after coloration (simple 1-layer Cauchy model). From Ref. [77].

X(cm)	k Amplitude (Error ± 0.005)
−3.5	0.0002
−3	0.0025
−2.5	0.044
−2	0.004
−1.5	0.015
−1	0.025
−0.5	0.056
0	0.041
0.5	0.092
1	0.105
1.5	0.075

Table 1.
k Amplitude vs. Position at the center line after the colorization in the dry state. From Ref. [77].

electrochromic efficiencies of mixed metal oxides (TiO_2-SnO_2) deposited by reactive sputtering were also mapped, too, see **Figure 5**.

The coloration process was followed in situ by SE at the center point of the mixed metal oxide – highly conductive Si sample. They could map the colorized layer using a simple one-layer Cauchy dispersion optical model after the coloration ±process. For the Cauchy model, the k Amplitude (extinction) parameter has been considered a good indicator of the CE, as it is shown in **Figure 5b**. The maximum k value exhibits that the optimal composition is at (30%) TiO_2– (70%) SnO_2. See **Table 1**.

4. Conclusions

Many binary oxides were studied as potentially promising EC materials. However, most of the studies have investigated only a few compositions. Some of them studied only the role of adding a single percentage of a secondary material, emphasis of research was put on studying the effects of doping. Only a few examples can be found where a comprehensive investigation spanning the full compositional range between the component oxides was made. Note, that in most cases the mixed metal oxides showed better EC properties than the pure oxides. For example, Ismaeel et al. [75] found a fourfold enhancement of CE in Titanium-Molybdenum oxide mixed films. Therefore, in order to enhance the EC performance of materials further focus should be put on studying properties of mixed oxide materials in the full (0–100%) composition range. Combinatorial sputtering techniques offer a feasible way to prepare samples for this purpose.

Acknowledgements

This work has been funded by NKFIH OTKA K 143216 and 146181 projects. Project TKP2021-EGA-04 acknowledges the support from the Ministry of Innovation and Technology of Hungary financed under the TKP2021 funding scheme. The work with

20FUN02 "POLight" project has received funding from the EMPIR program, from the European Union's Horizon 2020 research and innovation program. Noor Taha Ismaeel acknowledges the Stipendium Hungaricum scholarship.

Author details

Zoltán Lábadi[1], Noor Taha Ismaeel[2,3], Peter Petrik[1,4] and Miklós Fried[1,5]*

1 Institute of Technical Physics and Materials Science, Centre for Energy Research, Budapest, Hungary

2 Doctoral School on Materials Sciences and Technologies, Óbuda University, Budapest, Hungary

3 Institute of Laser for Postgraduate Studies, University of Baghdad, Baghdad, Iraq

4 Faculty of Science and Technology, Department of Electrical Engineering, Institute of Physics, University of Debrecen, Debrecen, Hungary

5 Institute of Microelectronics and Technology, Óbuda University, Budapest, Hungary

*Address all correspondence to: fried.miklos@uni-obuda.hu

References

[1] Granqvist CG. Oxide electrochromics: An introduction to devices and materials. Solar Energy Materials & Solar Cells. 2012;**99**:1-13. DOI: 10.1016/j.solmat.2011.08.021

[2] Lee JW, Jung HJ, Park JY, Lee JB, Yoon Y. Optimization of building window system in Asian regions by analyzing solar heat gain and daylighting elements. Renewable Energy. 2013;**50**:522-531. DOI: 10.1016/j.renene.2012.07.029

[3] Granqvist CG. Electrochromic materials: Out of a niche. Nature Materials. 2006;**5**:89-90. DOI: 10.1038/nmat1577

[4] Llordé SA, Garcia G, Gazquez J, Milliron DJ. Tunable near-infrared and visible-light transmittance in nanocrystal-in glass composites. Nature. 2013;**500**:323-326. DOI: 10.1038/nature12398

[5] Barile CJ, Slotcavage DJ, Hou J, Strand MT, Hernandez TS, McGehee MD. Dynamic windows with neutral color, high contrast, and excellent durability using reversible metal electrodeposition. Joule. 2017;**1**:133-145. DOI: 10.1016/j.joule.2017.06.001

[6] DeForest N, Shehabi A, O'Donnell J, Garcia G, Greenblatt J, Lee ES, et al. United States energy and CO_2 savings potential from deployment of near-infrared electrochromic window glazings. Building and Environment. 2015;**89**:107-117. DOI: 10.1016/j.buildenv.2015.02.021

[7] Cheng W, He J, Dettelbach KE, Johnson NJJ, Sherbo RS, Berlinguette CP. Photodeposited amorphous oxide films for Electrochromic windows. Chem. 2018;**4**:821-832. DOI: 10.1016/j.chempr.2017.12.030

[8] Gillaspie DT, Tenent RC, Dillon AC. Metal-oxide films for electrochromic applications: Present technology and future directions. Journal of Materials Chemistry. 2010;**20**:9585-9592. DOI: 10.1039/C0JM00604A

[9] Runnerstrom EL, Llordé SA, Lounis SD, Milliron DJ. Nanostructured electrochromic smart windows: Traditional materials and NIR-selective plasmonic nanocrystals. Chemical Communications. 2014;**50**:10555-10572. DOI: 10.1039/C4CC03109A

[10] Granqvist CG. Electrochromics for smart windows: Oxide-based thin films and devices. Thin Solid Films. 2014;**564**:1-38. DOI: 10.1016/j.tsf.2014.02.002

[11] Cannavale A, Cossari P, Eperon GE, Colella S, Fiorito F, Gigli G, et al. Forthcoming perspectives of photoelectrochromic devices: A critical review. Energy & Environmental Science. 2016;**9**:2682-2719. DOI: 10.1039/C6EE01514J

[12] Granqvist CG. Handbook of Inorganic Electrochromic Materials. Amsterdam, Netherlands: Elsevier; 1995. ISBN: 9780080532905

[13] González-Borrero PP, Sato F, Medina AN, Baesso ML, Bento AC, Baldissera G, et al. Optical band-gap determination of nanostructured WO3 film. Applied Physics Letters. 2010;**96**:061909. DOI: 10.1063/1.3313945

[14] Novinrooz A, Sharbatdaran M, Noorkojouri H. Structural and optical properties of WO3 electrochromic layers

prepared by the sol-gel method. Central European Journal of Physics. 2005;**3**: 456-466. DOI: 10.2478/BF02475650

[15] Hsu C-S, Chan C-C, Huang H-T, Peng C-H, Hsu W-C. Electrochromic properties of nanocrystalline MoO3 thin films. Thin Solid Films. 2008;**516**: 4839-4844. DOI: 10.1016/j.tsf.2007.09.019

[16] Chaichana S, Sikong L, Kooptarnond K, Chetpattananondh K. The electrochromic property of MoO3/WO3 nanocomposite films. IOP Conference Series: Materials Science and Engineering. 2018;**378**:012002. DOI: 10.1088/1757-899X/378/1/012002

[17] Colton RJ, Guzman AM, Rabalais JW. Photochromism and electrochromism in amorphous transition metal oxide films. Accounts of Chemical Research. 1978;**11**:170-176. DOI: 10.1021/ar50124a008

[18] Wen RT, Granqvist CG, Niklasson GA. Eliminating degradation and uncovering ion-trapping dynamics in electrochromic WO3 thin films. Nature Materials. 2015;**14**:996-1001. DOI: 10.1038/nmat4368

[19] Llordes A, Wang Y, Fernandez-Martinez A, Xiao P, Lee T, Poulain A, et al. Linear topology in amorphous metal oxide electrochromic networks obtained via low-temperature solution processing. Nature Materials. 2016;**15**: 1267-1273. DOI: 10.1038/nmat4734

[20] Lábadi Z, Takács D, Zolnai Z, Petrik P, Fried M. Compositional optimization of sputtered WO3/MoO3 films for high coloration efficiency. Materials. 2024;**17**:1000. DOI: 10.3390/ma17051000

[21] Sauvet K, Sauques L, Rougier A. IR electrochromic WO3 thin films: From optimization to devices. Solar Energy Materials & Solar Cells. 2009;**93**: 2045-2049. DOI: 10.1016/j.solmat.2009.05.003

[22] Sato R, Kawamura N, Tokumaru H. Relaxation mechanism of electrochromism of tungsten-oxide film for ultra-multilayer optical recording depending on sputtering conditions. Japanese Journal of Applied Physics. 2007;**46**:3958. DOI: 10.1143/JJAP.46.3958

[23] Liao C-C, Chen F-R, Kai J-J. Annealing effect on electrochromic properties of tungsten oxide nanowires. Solar Energy Materials & Solar Cells. 2007;**91**:1258. DOI: 10.1016/j.solmat.2007.04.014

[24] Barbosa PC, Silva MM, Smith MJ, Gonçalves A, Fortunato E. Optical devices performance with poly (trim ethylene carbonate) based electrolytes. Thin Solid Films. 2009;**516**:01-1480. DOI: 10.1149/MA2009-01/45/1486

[25] Beydaghyan G, Renaud J-L, Bader G, Ashrit PV. Enhanced electrochromic properties of heat-treated nanostructured tungsten trioxide thin films. Journal of Materials Research. 2008;**23**:274-280. DOI: 10.1557/JMR.2008.0037

[26] Joraid AA. Comparison of electrochromic amorphous and crystalline electron beam deposited WO3 thin films. Current Applied Physics. 2009;**9**:73-79. DOI: 10.1016/j.cap.2007.11.012

[27] Hari Krishna K, Hussain OM, Julien CM. Electrochromic properties of nanocrystalline WO3 thin films grown on flexible substrates by plasma-assisted evaporation technique. Applied Physics A: Materials Science & Processing. 2010;

99:921-929. DOI: 10.1007/s00339-010-5681-5

[28] Sauvet K, Rougier A, Sauques L. Electrochromic WO3 thin films active in the IR region. Solar Energy Materials & Solar Cells. 2008;**92**:209-215. DOI: 10.1016/j.solmat.2007.01.025

[29] Rougier A, Sauvet K, Sauques L. Electrochromic materials from the visible to the infrared region: An example WO3. Ionics. 2008;**14**:99-105. DOI: 10.1007/s11581-007-0191-y

[30] Deshpande R, Lee S-H, Mahan AH, Parilla PA, Jones KM, Norman AG, et al. Opti-mization of crystalline tungsten oxide nanoparticles for improved electrochromic applications. Solid State Ionics. 2007;**178**:895-900. DOI: 10.1016/j.ssi.2007.03.010

[31] Gubbala S, Thangala J, Sunkara MK. Nanowire-based electrochromic devices. Solar Energy Materials & Solar Cells. 2007;**91**:813-820. DOI: 10.1016/j.solmat.2007.01.016

[32] White CM, Gillaspie DT, Whitney E, Lee S-H, Dillon AC. Flexible electrochromic devices based on crystalline WO3 nanostructures produced with hot-wire chemical vapor deposition. Thin Solid Films. 2009;**517**:3596-3599. DOI: 10.1016/j.tsf.2009.01.033

[33] Bathe SR, Patil PS. Electrochromic characteristics of fibrous reticulated WO3 thin films prepared by pulsed spray pyrolysis technique. Solar Energy Materials and Solar Cells. 2007;**91**:1097-1101. DOI: 10.1016/j.solmat.2007.03.005

[34] Bathe SR, Patil PS. Titanium doping effects in electrochromic pulsed spray pyrolysed WO3 thin films. Solid State Ionics. 2008;**179**:314-323. DOI: 10.1016/j.ssi.2008.02.052

[35] Kadam PM, Tarwal NL, Shinde PS, Patil RS, Deshmukh HP, Patil PS. From beads-to-wires-to-fibers of tungsten oxide: Electrochromic response Appl. Physica A. 2009;**97**:323-330. DOI: 10.1007/s00339-009-5334-8

[36] Kim C-Y, Cho S-G, Park S, Choi D-K. The Journal of Ceramic Processing Research. 2009;**10**:851

[37] Bertus LM, Enesca A, Duta A. Influence of spray pyrolysis deposition parameters on the optoelectronic properties of WO3 thin films. Thin Solid Films. 2012;**520**:4282-4290. DOI: 10.1016/j.tsf.2012.02.052

[38] Šurca Vuk A, Jovanovski V, Pollet-Villard A, Jerman I, Orel B. Solar Energy Materials & Solar Cells. 2008;**92**:126-135. DOI: 10.1016/j.solmat.2007.01.023

[39] Balaji S, Djaoued Y, Albert A-S, Ferguson RZ, Brüning R. Hexagonal tungsten oxide based Electrochromic devices: Spectroscopic evidence for the Li ion occupancy of four-Coordinated Square windows. Chemistry of Materials. 2009;**21**:1381. DOI: 10.1021/cm8034455

[40] Balaji S, Djaoued Y, Albert A-S, Ferguson RZ, Brüning R, Su B-L. Construction and characterization of tunable me-so−/macroporous tungsten oxide-based transmissive electrochromic devices. Journal of Materials Science. 2009;**44**:6608-6616. DOI: 10.1007/s10853-009-3575-8

[41] Kondrachova LV, May RA, Cone CW, Vanden Bout DA, Stevenson KJ, Langmuir. Evaluation of lithium-ion insertion reactivity via Electrochromic diffraction-based imaging. Mesoporous Electrodes. 2009;**25**:2508-2518. DOI: 10.1021/la803245a], 10.1021/la803245a]

[42] Nah Y-C, Ghicov A, Kim D, Schmuki P. Self-organized nano-tubes of TiO_2 MoO_3 with enhanced electrochromic properties. Electrochemistry Communications. 2008;**10**:1777. DOI: 10.1039/b820953g

[43] Zhang J, Wang XL, Xia XH, Gu CD, Zhao ZJ, Tu JP. Enhanced electrochromic performance of macroporous WO3 films formed by anodic oxidation of DC-sputtered tungsten layers. Electrochimica Acta. 2010;**55**:6953-6958. DOI: 10.1016/j.electacta.2010.06.082

[44] Ou JZ, Balendhran S, Field MR, McCulloch DG, Zoolfakar AS, Rani RA, et al. The anodized crystalline WO_3 nanoporous network with enhanced electrochromic properties. Nanoscale. 2012;**4**:5980-5988. DOI: 10.1039/C2NR31203D

[45] Kang J-H, Paek S-M, Hwang S-J, Choy J-H. Optical iris application of electrochromic thin films. Electrochemistry Communications. 2008;**10**:1785-1787. DOI: 10.1016/j.elecom.2008.09.013

[46] Khoo E, Lee PS, Ma J. Electrophoretic deposition (EPD) of WO_3 nanorods for electrochromic application. Journal of the European Ceramic Society. 2010;**30**:1139-1144. DOI: 10.1016/j.jeurceramsoc.2009.05.014

[47] Green SV, Pehlivan E, Granqvist CG, Niklasson GA. Solar Energy Materials & Solar Cells. 2012;**99**:339-344. DOI: 10.1016/j.solmat.2011.12.025

[48] Sivakumar R, Gopinath CS, Jayachandran M, Sanjeeviraja C. An electrochromic device (ECD) cell characterization on elec-tron beam evaporated MoO3 films by intercalating/deintercalating the H+ ions. Current Applied Physics. 2007;**7**:76-86. DOI: 10.1016/j.cap.2005.12.001

[49] Patil RS, Uplane MD, Patil PS. Electrosynthesis of Electrochromic molybdenum oxide thin films with rod-like features. International Journal of Electrochemical Science. 2008;**3**:259-265. DOI: 10.1016/S1452-3981(23)15451-4

[50] Gesheva KA, Cziraki A, Ivanova T, Szekeres A. Crystallization of chemically vapor deposited molybdenum and mixed tungsten/molybdenum oxide films for electrochromic application. Thin Solid Films. 2007;**515**:4609-4613. DOI: 10.1016/j.tsf.2006.11.042

[51] Dhanasankar M, Purushothaman KK, Muralidharan G. Effect of tungsten on the electrochromic behavior of sol–gel dip coated molybdenum oxide thin films. Materials Research Bulletin. 2010;**45**:542-545. DOI: 10.1016/j.materresbull.2010.02.003

[52] Hsu C-S, Chan C-C, Huang H-T, Peng C-H, Hsu W-C. Electrochromic properties of nanocrystalline MoO_3 thin films. Thin Solid Films. 2008;**516**:4839-4844. DOI: 10.1016/j.tsf.2007.09.019

[53] Laurinavichute VK, Vassiliev SY, Plyasova LM, Molina IY, Khokhlov AA, Pugolovkin LV, et al. Cathodic electrocrystallization and electrochromic properties of doped rechargeable oxotungstates. Electrochimica Acta. 2009;**54**:5439-5448. DOI: 10.1016/j.electacta.2009.04.035

[54] Sorar I, Pehlivan E, Niklasson GA, Granqvist CG. Electrochromism of DC magnetron sputtered TiO_2 thin films: Role of deposition parameters. Solar Energy Materials & Solar Cells. 2013;**115**:172-180. DOI: 10.1016/j.solmat.2013.03.035

[55] Khalifa S, Lin H, Ismat Shah S. Structural and electrochromic properties of TiO_2 thin films prepared by metallorganic chem-ical vapor deposition. Thin Solid Films. 2010;**518**:5457-5462. DOI: 10.1016/j.tsf.2010.04.013

[56] Shinde PS, Deshmukh HP, Mujawar SH, Inamdar AI, Patil PS. Spray deposited titanium oxide thin films as passive counter electrodes. Electrochimica Acta. 2007;**52**:3114-3120. DOI: 10.1016/j.electacta.2006.09.053

[57] Zelakowska E, Rysiakiewicz-Pasek E. Thin TiO_2 films for an electrochromic system. Optical Materials. 2009;**31**: 1802. DOI: 10.1016/j.optmat.2008.12.037

[58] Lin S-Y, Chen Y-C, Wang C-M, Liu C-C. Journal of Solid State Electrochemistry. 2008;**12**:1481

[59] Wang C-M, Lin S-Y, Chen Y-C. Electrochromic properties of TiO_2 thin films prepared by chemical solution deposition method. Journal of Physics and Chemistry of Solids. 2008;**69**: 451-455. DOI: 10.1016/j.jpcs.2007.07.113

[60] Ivanova T, Harizanova A, Koutzarova T, Krins N, Vertruyen B. Electrochromic TiO_2, ZrO_2 and TiO_2–ZrO_2 thin films by dip-coating method. Materials Science and Engineering B. 2009;**165**:212-216. DOI: 10.1016/j.mseb.2009.07.013

[61] Hahn R, Ghikov A, Tsuchiya H, Macak JM, Muñoz AG, Schmuki P. Lithium-ion insertion in anodic TiO_2 nanotubes resulting in high electrochromic contrast. Physica Status Solidi (a). 2007;**204**:1281-1285. DOI: 10.1002/pssa.200674310

[62] Paramasivam I, Macak JM, Ghikov A, Schmuki P. Chemical Physics Letters. 2007;**445**:233

[63] Berger S, Ghicov A, Nah Y-C, Schmuki P. Transparent TiO_2 nanotube electrodes via thin layer Anodization: Fabrica-tion and use in Electrochromic devices. Langmuir. 2009;**25**:4841-4844. DOI: 10.1021/la9004399

[64] Cantao MP, Cisneros JI, Torresi RM. Electrochromic behavior of sputtered titanium oxide thin films. Thin Solid Film. 1995;**259**:70-74 SSDI 0040-6090 (94)06401-6

[65] Şilik E, Pat S, Özen S, Mohammadigharehbagh R, Yudar HH, Musaoğlu C, et al. Electrochromic properties of TiO_2 thin films grown by thermionic vacuum arc method. Thin Solid Film. 2017;**640**:27-32. DOI: 10.1016/j.tsf.2017.07.073

[66] Cai GF, Zhou D, Xiong QQ, Zhang JH, Wang XL, Gu CD, et al. Efficient electrochromic materials based on TiO_2- WO_3 core/shell nanorod arrays. Solar Energy Materials and Solar Cells. 2013;**117**:231-238. DOI: 10.1016/j.solmat.2013.05.049

[67] Patil PS, Mujawar SH, Inamdar AI, Sadale SB. Electrochromic properties of spray deposited TiO_2-doped WO_3 thin films. Applied Surface Science. 2005; **250**:117-123. DOI: 10.1016/j.apsusc.2004.12.042

[68] Dhandayuthapani T, Sivakumar R, Zheng D, Xu H, Ilangovan R, Sanjeeviraja C, et al. WO_3/TiO_2 hierarchical nanostructures for electrochromic applications. Materials Science in Semiconductor Processing. 2021;**123**:105515. DOI: 10.1016/j.mssp.2020.105515

[69] Ashok Reddy GV, Shaik H, Kumar KN, Madhavi V, Shetty HD, Sattar SA, et al. Structural and electrochemical studies of WO_3 coated TiO_2 nanorod hybrid thin films for

electrochromic applications. Optik. 2023;**277**:170694. DOI: 10.1016/j.ijleo.2023.170694

[70] Nah Y-C, Ghicov A, Kim D, Berger S, Schmuki P. TiO_2-WO_3 composite nanotubes by alloy Anodization: Growth and enhanced Electrochromic properties. Journal of the American Chemical Society. 2008; **130**(48):16154-16155. DOI: 10.1021/ja807106y

[71] Burdis MS, Siddle JR, Batchelor RA, Gallego JM. V0.50Ti0.50Ox thin films as counter electrodes for electrochromic devices. Solar Energy Materials and Solar Cells. 1998;**54**:93-98. DOI: 10.1016/S0927-0248(98)00059-2

[72] Marcel C, Brigouleix C, Vincent A, Plessis D, Nouhaud G, Hamon Y, et al. Electrochromic properties of lithium flexible devices based on tungsten and vanadium-titaniumoxide thin films. In: Rougier A, Rauh D, Nazri GA, editors. Electrochromic Materials and Applications. Vol. 27. Pennington, New Jersey, USA: The Electrochemical Society, Inc.; 2003. p. 218

[73] Nabeen KS, Nah Y-C, Tsuchiya H. Patrik Schmuki: Self-organized nano-tubes of TiO_2–MoO_3 with enhanced electrochromic properties. Chemical Communications. 2009;**15**:2008-2010. DOI: 10.1039/b820953g

[74] Ezhilmaran B, Bhat SV. Enhanced charge transfer in TiO_2 nanoparticles/ MoO_3 nanostructures bilayer heterojunction electrode for efficient electrochromism. Materials Today Communications. 2022;**31**:103497. DOI: 10.1016/j.mtcomm.2022.103497

[75] Ismaeel NT, Lábadi Z, Fried M. Investigation of Electrochromic Behavior of Combinatorial TiO_2-MoO_3 Mixed Layers. Available from: https://www.preprints.org/manuscript/202407.0422/v1, DOI:10.20944/preprints202407.0422.v1

[76] Habashyani S, Mobtakeri S, Budak HF, Kasapoğlu AE, Çoban Ö, Gür E. Electrochromic properties of undoped and Ti-doped MoO_3 converted from nano-wall MoS_2 thin films. Electrochimica Acta. 2024;**498**(10): 144638. DOI: 10.1016/j.electacta.2024.144638

[77] Ismaeel NT, Lábadi Z, Petrik P, Fried M. Investigation of Electrochromic, combinatorial TiO_2-SnO_2 mixed layers by spectroscopic Ellipsometry using different optical models. Materials. 2023;**16**(12):4204. DOI: 10.3390/ma16124204

Chapter 6

Perspective Chapter: Modification Engineering of Titanium Dioxide-Based Nanostructured Photocatalysts for Efficient Removal of Pollutants from Water

Martina Kocijan and Matejka Podlogar

Abstract

Titanium dioxide (TiO_2) is a semiconductor photocatalyst with remarkable attention due to its prospective environmental remediation applications. TiO_2's unique properties, such as photocorrosion resistance, chemical stability, and low toxicity, have prompted significant interest from worldwide researchers over the last decades. TiO_2, with its wide band gap of ~ 3.2 eV and a recombination rate of photo-induced charge carriers, possesses low quantum efficiency and photocatalytic activity when using the overall solar spectrum. To improve photocatalytic performance with the solar spectrum, it is necessary to strengthen the number of active sites on the material surface to promote its adsorption capacity, separation, and carrier transport. This chapter aims to give an overview of recent research work developed with TiO_2-based nanostructured photocatalysts to create high-throughput technologies for water treatment of a wide range of pollutants. Here, the novel engineering modifications of TiO_2-based photocatalyst nanostructures are summarized, and discussed. This review intends to provide robust information on the modification strategies of TiO_2-based nanostructured photocatalysts to remove persistent pollutants from water and develop sustainable environmental technologies.

Keywords: titanium dioxide, modification engineering, photocatalysis, persistent pollutant, wastewater remediation

1. Introduction

The most crucial component of life on Earth, water controls the well-being of the entire ecosystem. However, water quality is becoming increasingly endangered due to pollution from various organic and inorganic compounds resulting from growing urbanization and industrialization [1, 2]. The availability of clean water and energy sources is among the highest priorities for sustainable economic growth and humanity. In addition, the lack of drinking water is a global problem that can be solved by implementing a water conservation policy, the primary goal of which is the purification of wastewater and its reuse in industrial processes for irrigation or livestock feeding, which would save drinking water [3, 4]. The presence of pollutants in water has a harmful effect on the environment and human health. Therefore, using advanced oxidation processes, which imply the effective removal of contaminants, is necessary to preserve the environment and human health. Consequently, there is a request for effective and sustainable removal of pollutants from water because conventional methods of removing contaminants from wastewater are not sufficiently efficient, economical, and environmentally sustainable due to the use of enormous resources such as energy consumption, time, infrastructure, labor, and capital, more attention is giving to advanced water treatment processes as promising processes for the effective removal of persistent pollutants from wastewater without creating secondary pollutants [5]. Photochemistry has become a fundamental sustainable development goal for clean water and energy [6].

Advanced oxidation processes attract attention due to the possibility of completely removing many pollutants through oxidation and reduction reactions in water [7, 8]. The hydroxyl radical ($^{\bullet}OH$) is advanced oxidation's processes main reactive oxygen species. This radical can degrade various pollutants with its high reactivity and strong oxidation potential (E° = 2.8 V). $^{\bullet}OH$ is an extremely unstable radical with a short half-life of only 10^{-10} s. In terms of reactivity, it is right behind fluorine, which has an oxidation potential of E° = 3.03 V, which gives it the ability to degrade a wide range of pollutants with a high degradation rate constant of the order of 10^{6}–10^{9} M^{-1} s^{-1} [9, 10]. **Figure 1** shows the specific properties of hydroxyl radicals. Among them, photocatalysis based on semiconductors is an effective process in wastewater treatment due to its potential in purification wastewater where photons assist the reaction process with a catalyst [11]. The photocatalytic process is surface-driven, so its efficiency depends primarily on the photocatalyst's particle size, morphology, and specific surface area. Nowadays, nanostructured materials are most often used because they have a higher surface-to-volume ratio than their bulk materials, which are more compact [12].

Among the available catalysts, TiO_2 is a well-known photocatalyst among metal oxides due to its exceptional properties. TiO_2 belongs to n-type semiconductor

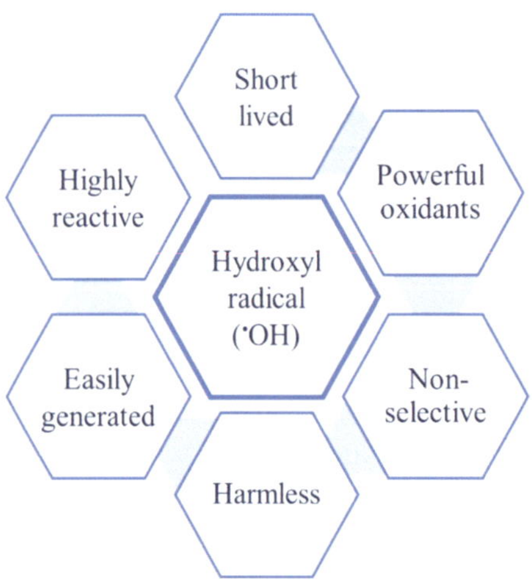

Figure 1.
Schematic illustration of hydroxyl radical properties.

materials. From 1972 until today, TiO_2 has been considered the most efficient semiconductor photocatalyst and one of the most frequently used benchmarks of standard photocatalysts in the field of environmental protection due to their excellent properties such as exceptional photocatalytic activity, non-toxicity, good mechanical, chemical, thermal stability, and low price [13, 14]. Semiconductor nanomaterial TiO_2 has a wide range of applications in areas such as photocatalysis (hydrogen production, wastewater treatment, and disinfection), pigments, solar cells, medicine, especially cancer therapy, sensors, cosmetics, electrodes for photovoltaic cells, protection of metals from corrosion and antibacterial surfaces which are activated by light [15, 16]. TiO_2 has unfavorable charge carrier dynamics and a wide band gap energy (~3.2 eV for the anatase and brookite phase and ~3.0 eV for the rutile phase), and modification of TiO_2 is required to improve the separation of charge carriers and the absorption of a broader spectrum of light for more efficient photocatalytic activity [17, 18]. To overcome the lack of TiO_2 and improve solar energy use as a green energy source, extensive research is being done to develop new methods to modify TiO_2 nanostructures. Modifying the optical and physicochemical properties of TiO_2 nanostructure is imperative to use a broader spectrum of solar illumination, increasing photocatalytic efficiency and saving energy. Modified TiO_2-based nanostructured photocatalysts significantly impact the narrowing of the band gap energy and the configuration of the surface structure, which provides a significant quant efficiency and reaction rates for the decomposition of organic and inorganic pollutants from water under solar illumination [19, 20].

2. Modification strategies for enhanced photoactivity of TiO_2 under solar irradiation

The preparation methods are essential in synthesizing TiO_2-based nanostructured photocatalysts with extraordinary photocatalytic properties. The synthesis pathways for TiO_2-based nanostructured photocatalysts are diverse, and the intention is to enhance the photocatalytic performance of these materials. Hence, the development of TiO_2-based nanostructured photocatalyst is an innovative approach to improve the application of nanomaterials in solar photocatalysis.

Principal preparation methods of TiO_2-based nanostructured photocatalysts can be classified as (i) *in-situ* and (ii) *ex-situ* crystallization [21, 22]:

In-situ crystallization involves synthesizing an additional component directly on a substrate or precursor material through reaction preparation steps. Furthermore, the functional groups on the substrate can control the size, morphology, and crystallinity of the grown semiconductor crystals. Additionally, the direct contact of the substrate with semiconductor nanoparticles promotes electron transfer [22].

In the *ex-situ* crystallization method, two or more components are mixed and combined using various preparation techniques. This method allows for the co-assembly of nanoparticles with different textures, sizes, and shapes on substrates. However, the *ex-situ* approach is limited by nonuniform distribution and low density of semiconductors on the substrate surface [22].

Accordingly, the modification strategies that are adopted to improve and enhance the light absorption, charge separation, and TiO_2 surface reactivity are nonmetal (N, C, P, S, B) and metal (Ag, Sn, Au, Fe, Pd, Cu) doping [19–21], defect structures (oxygen vacancies and Ti^{3+}) in the crystal lattice [22], chemical coupling with other semiconductors (WO_3, SnO_2, ZnO, ZnS, CdS, CeO_2, Fe_2O_3) [23, 24], and carbon-based materials (graphene, GO, rGO, g-C_3N_4) [25–28].

Several well-known synthesis methods are most utilized to prepare TiO_2-based nanostructured photocatalysts. The hydrothermal method is commonly used for *in situ* synthesis [21]. High pressure and temperature during hydrothermal processes promote strong interactions between compounds, forming solid interfacial bonds. Hydrothermal technique has been increasingly accepted for synthesizing TiO_2-based nanostructured since it provides advantages compared with other synthesis processes, such as low energy consumption, high reactivity, mild reaction conditions, simple set-up, and solvent control. Reaction synthesis is sensitive to experimental conditions, such as reaction temperature and time. However, it can still produce high efficiency TiO_2-based nanostructured photocatalysts with effective properties at a low cost [23]. Maletić et al. conducted a synthesis reaction at 160°C for 12 h, combining a glucose solution with hydrochloric acid and titanium isopropoxide. This result obtained is a TiO_2/HTC composite with higher porosity, firmly linked through Ti-O-C chemical bonds [24]. Sahoo et al. synthesized the pure rutile phase of TiO_2 nanorods with microflower morphology using a hydrothermal method and decorated them with Au to enhance photocatalytic degradation efficiency [25]. Further, a defect-oriented hydrothermal approach synthesizes manganese-doped titanium dioxide nanoparticles ($MnTiO_2$-NPs) [26].

The sol-gel approach is a wet chemical method for synthesizing TiO_2-based nanostructures. It is based on the phase transition of a reaction from a colloidal liquid "sol" into a solid gel through hydrolysis and polycondensation reaction [21, 27]. Ma et al. adopted a sol-gel approach to fabricate N-TiO_2/RGO nanocomposites. They added tetrabutyl titanate and acetylacetone into ethyl alcohol under vigorous magnetic stirring. After stirring the mixture for 2 h at room temperature, a nanocolloid (sol-TiO_2) was obtained. The urea solution was added to the nanocolloid sol-TiO_2 under vigorous magnetic stirring for 2 h to form nitrogen-doped gel-TiO_2. Then, a solution of GO was added to the nitrogen-doped gel-TiO_2, with a mixture of nitric acid and ethyl alcohol added to avoid destroying the stable gel system. The mixture was stirred for 1.5 h and then dried at 80°C, and dried powders were calcined in a nitrogen atmosphere for 1 h [28]. The sol-gel route's advantages include a low process temperature, the ability to control the molecular-scale composition, and the final product's homogeneity and high purity. However, a significant drawback is that the synthesis product is in an amorphous rather than a crystal phase and, thus, requires an annealing step for crystallization [29].

Recently, the microwave-assisted method for nanoparticle synthesis has gained popularity because it involves homogeneity and rapid heating of the reaction mixture at the required temperature, leading to time savings and producing smaller and more uniform particles than other *in situ* approaches. Still, microwave-assisted methods require high-power microwave heating, with high energy consumption. Also, the microwave-assisted approach does not allow for monitoring the growth of nanoparticles over synthesis time or for the preparation of large quantities of photocatalysts [30]. Hardiansyah et al. reported a microwave-assisted synthesis of reduced graphene oxide/titanium dioxide nanocomposites. In the experiment procedure, water-dispersed GO and TiO_2 (anatase) powder were reduced simultaneously during microwave irradiation [31].

The *ex-situ* crystallization method involves mixing commercially available or already synthesized nanomaterials with colloidal suspension of graphene-based templates, where nanoparticles attach to the graphene materials through oxygen functional groups [21]. The mechanochemical method for nanoparticle preparation is obtained without using organic solvents or high temperatures. Therefore, it could reduce the negative impact on the environment. Nevertheless, high energy ball milling may require high energy consumption [32]. Disadvantages of mechanochemical synthesis compared to other methods include the susceptibility of the synthesis products to contamination from the production of by-products, milling balls, or milling reactors. Synthesis products may contain low crystallinity due to defects caused by mechanical energy input. When preparing highly soluble metal oxides that decompose after contact with washing liquids, they cannot be easily separated from the synthesis by-products [33]. González et al. reported the synthesis of a few layers of graphene (FLG) decorated with TiO_2 nanoparticles-hybrid nanocomposites using a mechanochemical approach [34].

A synthesis technique can modify the morphological and optical properties of TiO_2-based nanostructured photocatalysts, and it is imperative to find the optimal synthesis strategy for preparing TiO_2-based nanostructured photocatalysts to remove persistent pollutants from water efficiently. However, different photocatalytic parameters, such as the intensity of light irradiation, pH of the solution, the structure of the contaminants, water environments, etc., can also affect the photocatalytic efficiency, and it is necessary to investigate those parameters.

3. Application of TiO_2-based nanostructured photocatalysts for wastewater remediation

Organic contaminants, a persistent issue in textile, pharmaceutical, and agricultural wastewater, as well as in groundwater, surface water, and municipal sewage, are a cause for concern due to their potential adverse environmental effects over time [35, 36]. TiO_2-based materials suggest a promising solution for the efficient photocatalytic degradation of contaminants. Possible benefits include strong interaction between materials, accelerating charge migration, large specific surface area, and boosting the adsorption of pollutants. However, up till now, research offers hope for its practical application. This chapter presents the latest TiO_2-based nanostructured photocatalysts, which generally exhibit superior photocatalytic performances compared to pure TiO_2.

The new structure's intense interaction between materials will enhance its chemical and physical properties and the separation and migration of photoinduced charge. This will improve the lifetime of charge carriers and reduce recombination. Moreover, TiO_2-based nanostructured photocatalysts can increase the production of reactive oxygen species, thus improving the photocatalytic activity. Given that the photocatalytic oxidation process primarily occurs on the exterior of the photocatalyst, a superior surface-based adsorption performance is a prerequisite for efficient contaminant decomposition. The catalyst surface (oxygen functional groups) is essential for adsorption. Their presence significantly enhances the ability to adsorb and decompose more pollutants in wastewater, particularly near the photocatalyst surface. This, in turn, leads to high photocatalytic efficiency, a crucial factor in the successful application of photocatalysts in wastewater treatment.

The oxidation rate and the photocatalysis reaction's efficiency for pollutant degradation depend on various process factors. It was observed that the photocatalytic activity increases with a decrease in the particle size, which results in a higher surface-to-volume ratio and, consequently, a higher surface photoactivity [37, 38]. However, some studies prove that particle size and surface area do not have a direct effect on the photocatalytic activity of the photocatalyst but that the photocatalytic activity is more influenced by surface properties such as the isoelectric point and pH value of the solution [39, 40]. In addition, synthesis parameters such as the selected precursor, synthesis temperature, time, and heat treatment temperature can significantly affect the photocatalytic activity [41, 42].

The surface morphology of the catalyst is a crucial factor in its use as a photocatalyst; since all chemical processes take place on the surface, an attempt is made to increase the specific surface area of photocatalysts. The optimal amount of each material in the nanocomposite plays a crucial role in obtaining a large specific surface area of the photocatalyst [43]. An excessive amount of the added material in the nanocomposite prevents light from reaching the TiO_2, limiting the photocatalyst's activity. Suppose the number of components in the nanocomposite is too small. In that case, TiO_2 nanoparticles create clusters on the surface, of which only a tiny number of TiO_2 nanoparticles interact with the added material, forming a bond between the materials, which limits electron transfer [44]. The nanocomposite with the optimal ratio of the number of components in the nanocomposite has the highest rate of electron charge transfer. Consequently, it improves the degradation of pollutants from the aqueous medium. Fu et al. investigated the optimal amount of GO (3–28 wt%) in GO/TiO_2 composite photocatalyst [45]. It was found that the best photodegradation rate of methylene blue (MB) reached 91% using 15 wt% of GO in the GO/TiO_2 nanocomposite. Yadav et al. reported the degradation of rhodamine B by using a type II heterojunction of titanium dioxide/graphitic carbon nitride (TiO_2/*g*-C_3N_4) photocatalyst. The synthesized TiO_2/*g*-C_3N_4 nanocomposite (5 wt% *g*-C_3N_4) showed enhanced photocatalytic activity for the rhodamine B dye degradation compared to pristine TiO_2 and *g*-C_3N_4 nanoparticles. This was attributed to the synergetic effect between TiO_2 and *g*-C_3N_4 in the synthesized type II heterojunction TiO_2/*g*-C_3N_4 nanocomposite [46]. Najafi et al. prepared TiO_2-GO nanocomposites with nanoparticle and nanowire morphologies of TiO_2 using a hydrothermal route. In the tetragonal structure of TiO_2-GO nanocomposite, covalent bonds were established between GO sheets and TiO_2 nanostructures. The highest removal rate of MB was found using nanowire TiO_2-GO nanocomposites [47].

The amount of catalyst in the photocatalytic process for the decomposition of pollutants from the aqueous medium affects the overall rate of the photocatalytic reaction, where the amount of catalyst is directly proportional to the overall rate of the photocatalytic reaction. By increasing the amount of catalyst, the specific surface area of the catalyst increases; that is, the number of active sites on the surface of the photocatalyst increases. Consequently, more reactive radicals are produced for the pollutant decomposition reaction [48]. Linear dependence exists to a certain degree when the reaction rate begins to deteriorate and becomes independent of the added amount of catalyst. However, the linear relationship ceases to be valid if the dose of photocatalyst is above the optimal amount, that is, in excess, because the pollutant decomposition reaction slows down. High turbidity occurs when the amount of catalyst increases above the saturation level. This process can be explained by the fact that the radiation can no longer penetrate the solution due to photocatalyst agglomeration. As a result, photocatalyst agglomerates hinder and limit the penetration of light into the solution on a sizable catalytic area; that is, more radiation will be scattered, which leads to a decrease in the efficiency of the photocatalytic reaction and the percentage of pollutant degradation will decrease [49].

The decomposition rate of pollutants depends on the pH of the solution that controls the adsorption of the pollutant, that is, the organic compound on the surface of the photocatalyst, since pollutant degradation reactions depend on the formation of radical species but also the electrostatic interaction between pollutants and photocatalysts [50]. Different photocatalysts have different points of zero charge (PZC), and the surface charges depend on the pH value. The PZC of the photocatalyst is defined as the threshold pH when the total surface charges of the catalyst are equal to zero [51]. Pollutants in the aqueous medium can be cationic, anionic, or amphoteric, and depending on the molecular interaction, the reaction will either speed up or slow down [52]. The point of zero charge (PZC) is formed where the zeta potential is zero and the catalyst has no charge. As a result, photocatalyst agglomerates are formed, which reduces the active surface between the organic pollutant and the photocatalyst surface, and the formed agglomerates inhibit the penetration of light, which leads to a decrease in the rate of dye decomposition [53, 54]. The surface of the photocatalyst will be negatively charged when $pH > pH_{pzc}$, positively charged when $pH < pH_{pzc}$, and neutral when $pH \approx pH_{pzc}$. Different pH values in the solution modify the electrical double-layer photocatalyst, consisting of a charged photocatalyst surface and pollutant molecules [55]. A schematic representation of the absorption of cationic and anionic pollutant molecules on the photocatalyst surface in acidic and alkaline conditions is shown in **Figure 2**. For example, Deshmukh et al. studied the degradation of methylene blue from an aqueous solution at different pH values (2.3, 4.0, 12.1, and 13.2) [50]. At lower pH values, that is, under acidic conditions, the surface of the TiO_2@rGO photocatalyst acquires a positive charge, leading to the active sites' protonation. As a cationic dye, MB creates electrostatic repulsion with positively charged dye molecules and the surface of the TiO_2@rGO photocatalyst. At higher pH values, the TiO_2@rGO photocatalyst surface becomes alkaline, leading to deprotonation. Under alkaline conditions, there is a more significant number of negative sites on the outer surface of the TiO_2@rGO photocatalyst, and complex formation occurs due to the interactions of dye cations with the negative sites of the TiO_2@rGO photocatalyst. Electrostatic attraction of dye molecules with TiO_2@rGO photocatalyst is significant at higher pH values. It was determined that PZC = 2.62 for TiO_2@rGO photocatalyst. The PZC value confirms the presence of a negative charge on the surface of the TiO_2@rGO photocatalyst at a higher pH value. In this case, the hydroxyl ions interact with the holes, which results in an increased formation of radicals and a decrease in the recombination of electron-hole pairs. The consequence is faster degradation of MB dyes at higher pH values, that is, in a basic medium. Chang et al. investigated the influence of initial pH values (2, 3, 4, 6, and 8) for phenol degradation in the presence of TiO_2/Fe_3O_4. The PZC of Fe_3O_4/TiO_2 was around pH = 10.9. Therefore, pH 2 is conducive to phenol degradation [53].

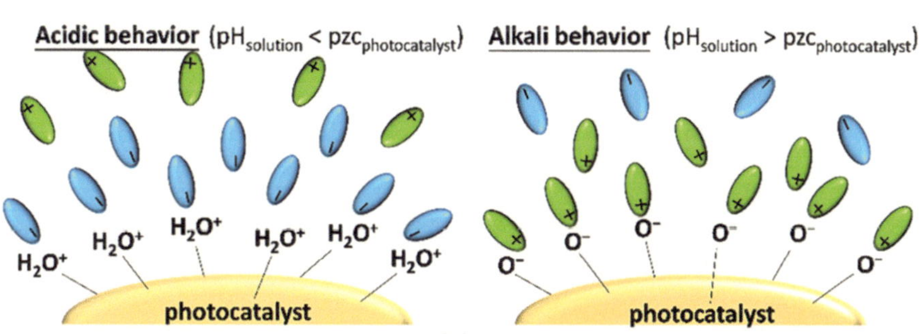

Figure 2.
Schematic illustration of the absorption of cationic and anionic dye molecules on the photocatalyst surface under acidic (left) and alkaline (right) conditions. Reprinted with permission from Ref. [56].

From a practical point of view, it is essential to investigate the pollutant removal efficiency as a function of the initial pollutant concentration because, in natural wastewater, the pollutant concentration continuously varies [57]. Different initial pollutant concentrations require different irradiation times to achieve a certain degree of removal of pollutants from the aqueous medium due to the ratio between the active, reactive sites on the surface of the photocatalyst and the polluted molecules. At a low initial concentration of pollutants, the number of reactive radicals on the surface of the photocatalyst is greater than the number of pollutants. The absence of pollutant molecules reduces the number of adsorbed molecules on the surface of the reacting photocatalyst with free radicals. By increasing the dye concentration, a more significant amount of pollutant molecules adsorbs on the surface of the photocatalyst. As a result, fewer photons are available and reach the catalyst's surface, resulting in less radical formation, that is, photocatalyst deactivation occurs, resulting in a lower percentage of pollutant decomposition from aqueous solutions. Therefore, the initial concentration of the pollutant should be optimized with the number of active, reactive sites, increasing the possibility of collision between the organic pollutant and the corresponding radical species to achieve the fastest removal of the pollutant from the aqueous media [57–59].

Light radiation (wavelength and light intensity) significantly impacts the photocatalytic process in the degradation of pollutants from the aqueous medium [60]. The photocatalytic activity of semiconductor oxides depends on the intensity of light absorption. The wavelength of light is related to Eg and the photon energy of the energy gap. Suppose the photon energy is too small, that is, smaller than the energy gap of the catalyst (Eg). In that case, the electrons will not be excited, and the surface's oxidation process will not occur [61]. Better photocatalytic efficiency can be achieved by increasing the light intensity to a specific value because it improves the charge separation of electrons and holes. Artificial radiation is more favorable than sunlight and can increase efficiency in degrading pollutants from water. Photocatalytic processes can also be carried out regardless of weather conditions. However, solar energy is emerging as an alternative and economical light source due to its abundance and safety. Solar photocatalytic reactions of pollutant degradation are carried out using direct sunlight. The main disadvantage of the solar photocatalytic reaction is the weather conditions required for the process's implementation [62–64].

Ollis summarized the effect of light intensity on the kinetics of photocatalytic degradation of dyes [65]. At low radiation intensities (0–20 mW cm^{-2}), the degradation rate increases linearly with increasing light intensity, while at medium light intensities (25 mW cm^{-2}), the rate depends on the square root of the light intensity. The speed is independent of the light intensity at high light intensities because there are more photons per unit of time and area. Thus, they increase the possibility of photon activation on the catalyst's surface, making the photocatalytic process more efficient. However, as the light intensity increases, the number of activation sites remains the same so that the reaction rate only reaches a certain level, even as the light intensity continues to increase [64, 66]. Pirgholi-Givi et al. investigated the influence of irradiation intensity and stirring rate on the photocatalytic activity of titanium dioxide nanostructures prepared by the microwave-assisted method for photodegradation of MB from water [67]. The photocatalytic activity of TiO_2 nanostructures increasing light intensity increases. It used a UV and Hg lamp located 15 cm above the surface of the reaction process. The stirring rate significantly affects the photocatalytic activity of TiO_2 nanostructures at low-intensity radiation. On the other hand, at high-intensity radiation, the stirring rate has a negligible influence on the photodegradation of MB.

Kocijan et al. monitored the photocatalytic degradation of MB from aqueous solution under simulated solar, UV-A, and natural sunlight irradiation using 32% g-C_3N_4@TiO_2 nanocomposite [68]. It was found that the MB removal depends on the light source and its intensity. The MB degradation rates decreased in the order of natural sunlight > simulated solar > UV-A irradiation.

In the literature, most of the photocatalytic degradation of organic pollutants was carried out with distilled or ultrapure water, aqueous media containing a minimal number of ions that do not inhibit the photocatalytic process, and the degradation of pollutants is very efficient. Real water media have more ions, with higher conductivity than ultrapure water [69]. Various ions in the aqueous medium can negatively or positively affect the photocatalytic efficiency of pollutant degradation [70]. From a practical point of view, the efficiency of the photocatalytic process should be tested in different natural water matrixes such as tap water, river water, lake water, seawater, industrial wastewater, etc. [43]. Natural waters usually contain inorganic salts as well as other organic substances. The water quality influences the degradation rate of pollutants; that is, the ions present can slow down or speed up the photocatalytic reaction [71]. The photocatalytic effect of pollutant removal from different natural water bodies should be studied before optimizing the process for commercial application. Kocijan et al. investigated the photocatalytic degradation of MB dye in different water matrixes such as ultrapure, river, lake, and seawater using 32% g-C_3N_4@TiO_2 nanocomposite [68]. The MB degradation rate was above 95% in all water matrixes under simulated solar irradiation. The degradation rate decreased in the order of ultrapure water > river water > lake water > seawater. A slight inhibition of the MB degradation rate in natural water bodies influences the presence of different ions, which act as scavengers in the photocatalytic process. Further, Kocijan et al. monitored MB degradation in natural water bodies using TiO_2@rGO_8 wt% nanocomposite under simulated solar irradiation [72]. The removal efficiency decreases in the order of tap water (95.98%) > ultrapure water (86.13%) > river water (78.07%) > lake water (67.48%) > seawater (56.22%), showing a significant dependence on the water medium used.

Photocatalytic drinking water disinfection is commonly performed by an ultraviolet light source that provides only 5% of the solar energy, making the process costly. To address this issue, functionalized nanostructured TiO_2-based photocatalysts are currently being developed, which can take advantage of visible light, speeding up the disinfection procedure cost-effectively. The unique physicochemical properties of TiO_2 semiconductors can be beneficial for disinfection applications. Graphene also plays a significant role in the photocatalytic inactivation of pathogen cells. Using graphene materials, particularly reduced graphene oxide in composites with TiO_2 nanoparticles, operate as a photogenerated electron transporter and acceptor, successfully suppressing the recombination of the photoexcited pairs, thus increasing the quantum efficiency of the microbial inactivation process [73].

Ch-Th et al. proposed a cost-effective and eco-friendly method to prepare GO/TiO_2 nanocrystals with high efficiency in deactivating *Escherichia coli* (*E. coli*) K12 under simulated visible light exposure. The study demonstrated that a small amount of the prepared photocatalyst achieved an impressive 99.9% deactivation of *E. coli* K12 in only 30 minutes. The research also emphasized the important role of the interaction between GO and hydroxyl ($^{\bullet}OH$)/superoxide ($O_2^{\bullet-}$) radicals from TiO_2 in the disinfection process [74].

Zhou et al., tested the antibacterial activity of the rGO-modified TiO_2 catalysts against *Enterobacter hormaechei* under UV-visible light illumination. The adsorption and the lower recombination of photoinduced electron-hole pairs in the prepared photocatalyst improved charge separation, which enhanced the antibacterial process. The survival ratio of *Enterobacter hormaechei* was reduced to about 10% within 90 minutes under visible light illumination. Structural and morphological characterization showed physical contact and disruptive interaction between prepared rGO-TiO_2 catalyst and *Enterobacter hormaechei*, mostly on rGO, which oxidized a cellular structure. Moreover, it was found that hydroxyl radicals were the main reactive oxygen species in the antibacterial process's mechanism [75].

Wanag et al. found that TiO_2 modified with reduced graphene oxide could effectively be utilized against *E. coli* bacteria's antibacterial activity under artificial solar light. Notably, the *E. coli* inactivation process is strongly related to the amount of reduced graphene oxide. The complete inactivation of *E. coli* was noticed after 75 minutes of illumination [76].

In summary, it is demonstrated that the efficient photocatalytic degradation rate of various organic pollutants and inactivating pathogens in water using photocatalysis depends on factors such as the extended light absorption range, separation of photogenerated charge carriers, the concentration of rGO in the nanocomposite, and formation of ROS.

4. Conclusion

Due to reduced potable water, water purification is imperative for future generations. The remarkable properties of titanium dioxide semiconductor material and simple synthesis approaches make titanium dioxide materials acceptable for creating diverse nanocomposites. Titanium dioxide-based nanostructured composites are acceptable for diverse engineering applications. The presented photocatalysts indicate the trend of preparing TiO_2-based materials to achieve synergetic effects of nanocomposites that typically exhibit enhanced structural, morphological, optical, and photocatalytic properties for the newly prepared nanocomposites compared to the pure TiO_2 nanoparticles. This approach has been widely accepted and utilized in solar-driven photocatalytic applications for water purification. Improved or novel synthesis methods, especially green ones, should be developed to conduct TiO_2-based nanocomposite photocatalysts to their full application potential. This could enable environmentally and cost-effective acceptable upscaling of laboratory synthesis, which could then be monitored in real applications for water purification and potentially contribute to society's sustainable development.

Conflict of interest

The authors declare no conflict of interest.

DOI: http://dx.doi.org/10.5772/intechopen.1007375

Author details

Martina Kocijan[1]* and Matejka Podlogar[2]

1 Department of Materials, Faculty of Mechanical Engineering and Naval Architecture, University of Zagreb, Zagreb, Croatia

2 Department for Nanostructured Materials, Jožef Stefan Institute, Ljubljana, Slovenia

*Address all correspondence to: martina.kocijan@fsb.unizg.hr

References

[1] Musie W, Gonfa G. Fresh water resource, scarcity, water salinity challenges and possible remedies: A review. Heliyon. 2023;**9**(8):e18685. Available from: https://www.sciencedirect.com/science/article/pii/S2405844023058930

[2] Cosgrove WJ, Loucks DP. Water management: Current and future challenges and research directions. Water Resources Research. 2015;**51**(6):4823-4839. DOI: 10.1002/2014WR016869

[3] Silva JA. Wastewater Treatment and Reuse for Sustainable Water Resources Management: A Systematic Literature Review, Sustainability. Vol. 15. Switzerland: Multidisciplinary Digital Publishing Institute (MDPI); 2023

[4] Intergovernmental Panel on Climate Change. Water cycle changes. In: Climate Change 2021 – The Physical Science Basis. Cambridge: Cambridge University Press; 2023. pp. 1055-1210

[5] Yeoh JX, Siti Nurul SNA, Syukri F, Koyama M, Nourouzi MM. Comparison between Conventional Treatment Processes and Advanced Oxidation Processes in Treating Slaughterhouse Wastewater: A Review, Water. Vol. 14. Switzerland: MDPI; 2022

[6] Watanabe K. Photochemistry on nanoparticles. In: Wandelt K, editor. Encyclopedia of Interfacial Chemistry. Oxford: Elsevier; 2018. pp. 563-572. Available from: https://www.sciencedirect.com/science/article/pii/B9780124095472132111

[7] Wang J, Zhuan R. Degradation of antibiotics by advanced oxidation processes: An overview. Science of the Total Environment. 2020;**701**:135023. Available from: https://www.sciencedirect.com/science/article/pii/S0048969719350156

[8] Coha M, Farinelli G, Tiraferri A, Minella M, Vione D. Advanced oxidation processes in the removal of organic substances from produced water: Potential, configurations, and research needs. Chemical Engineering Journal. 2021;**414**:128668. Available from: https://www.sciencedirect.com/science/article/pii/S1385894721002667

[9] Humayun M, Raziq F, Khan A, Luo W. Modification strategies of TiO_2 for potential applications in photocatalysis: A critical review. Green Chemistry Letters and Reviews. Vol. 11. Milton Park, Oxfordshire: Taylor and Francis Ltd; 2018. pp. 86-102

[10] Guerra-Rodríguez S, Rodríguez E, Singh DN, Rodríguez-Chueca J. Assessment of sulfate radical-based advanced oxidation processes for water and wastewater treatment: A review, Water. Vol. 10. Switzerland: MDPI AG; 2018

[11] Hong J, Cho KH, Presser V, Su X. Recent advances in wastewater treatment using semiconductor photocatalysts. Current Opinion in Green and Sustainable Chemistry. 2022;**36**:100644. Available from: https://www.sciencedirect.com/science/article/pii/S2452223622000566

[12] Seleš P, Vengust D, Radošević T, Kocijan M, Einfalt L, Kurtjak M, et al. Altering defect population during the solvothermal growth of ZnO nanorods for photocatalytic applications. Ceramics International. 2024;**50**(15):26819-26828. Available from: https://www.sciencedirect.com/science/article/pii/S0272884224018108

[13] Tomás-Gamasa M, Mascareñas JL. TiO_2-based photocatalysis at the interface with biology and biomedicine. Chembiochem. 2020;**21**(3):294-309. DOI: 10.1002/cbic.201900229

[14] Kang X, Liu S, Dai Z, He Y, Song X, Tan Z. Titanium dioxide: From engineering to applications. Catalysts. Vol. 9. Switzerland: MDPI; 2019

[15] Wang J, Wang Z, Wang W, Wang Y, Hu X, Liu J, et al. Synthesis, modification and application of titanium dioxide nanoparticles: A review. Nanoscale. 2022;**14**(18):6709-6734. DOI: 10.1039/D1NR08349J

[16] Ziental D, Czarczynska-Goslinska B, Mlynarczyk DT, Glowacka-Sobotta A, Stanisz B, Goslinski T, et al. Titanium dioxide nanoparticles: Prospects and applications in medicine. Nanomaterials. Vol. 10. Switzerland: MDPI AG; 2020

[17] Lettieri S, Pavone M, Fioravanti A, Amato LS, Maddalena P. Charge carrier processes and optical properties in TiO_2 and TiO_2-based heterojunction photocatalysts: A review. Materials. 2021;**14**(7):1645

[18] Eddy DR, Permana MD, Sakti LK, Sheha GAN, Solihudin GAN, Hidayat S, et al. Heterophase polymorph of TiO_2 (anatase, rutile, brookite, TiO_2 (B)) for efficient photocatalyst: Fabrication and activity. Nanomaterials. Vol. 13. Switzerland: MDPI; 2023

[19] Khan H, Shah MUH. Modification strategies of TiO_2 based photocatalysts for enhanced visible light activity and energy storage ability: A review. Journal of Environmental Chemical Engineering. 2023;**11**(6):111532. Available from: https://www.sciencedirect.com/science/article/pii/S2213343723022716

[20] Kocijan M. Razvoj fotokatalitičkoga nanokompozita na bazi titanijeva dioksida i reduciranoga grafenova oksida [doctoral dissertation]. Croatia: Faculty of Mechanical Engineering and Naval Architecture, University of Zagreb; 2023

[21] Chowdhury S, Balasubramanian R. Graphene/semiconductor nanocomposites (GSNs) for heterogeneous photocatalytic decolorization of wastewaters contaminated with synthetic dyes: A review. Applied Catalysis. B, Environmental. 2014;**160-161**:307-324. Available from: https://www.sciencedirect.com/science/article/pii/S0926337314003178

[22] Giovannetti R, Rommozzi E, Zannotti M, D'Amato CA. Recent advances in graphene based TiO_2 nanocomposites ($GTiO_2Ns$) for photocatalytic degradation of synthetic dyes. Catalysts. Vol. 7. Switzerland: MDPI; 2017

[23] Wong CL, Tan YN, Mohamed AR. A review on the formation of titania nanotube photocatalysts by hydrothermal treatment. Journal of Environmental Management. 2011;**92**(7):1669-1680. Available from: https://www.sciencedirect.com/science/article/pii/S0301479711000727

[24] Maletić M, Vukčević M, Kalijadis A, Janković-Častvan I, Dapčević A, Laušević Z, et al. Hydrothermal synthesis of TiO_2/carbon composites and their application for removal of organic pollutants. Arabian Journal of Chemistry. 2019;**12**(8):4388-4397. Available from: https://www.sciencedirect.com/science/article/pii/S1878535216300983

[25] Sahoo S, Kumar Sahu B, Shukla S, Srivastava SK, Sahoo PK. In-situ monitoring of plasmon-induced nanoscale photocatalytic activity from Au-decorated TiO_2 microflowers.

Nano Futures. 2023;**7**(2):025002. DOI: 10.1088/2399-1984/accf54

[26] Awan AM, Khalid A, Ahmad P, Alharthi AI, Farooq M, Khan A, et al. Defects oriented hydrothermal synthesis of TiO_2 and $MnTiO_2$ nanoparticles as photocatalysts for wastewater treatment and antibacterial applications. Heliyon. 2024;**10**(3):25579. DOI: 10.1016/j.heliyon.2024.e25579

[27] Bokov D, Turki Jalil A, Chupradit S, Suksatan W, Javed Ansari M, Shewael IH, et al. Nanomaterial by sol-gel method: Synthesis and application. Advances in Materials Science and Engineering. 2021;**2021**(1):5102014. DOI: 10.1155/2021/5102014

[28] Ma Y, Wang S, Zheng W, Xue X, Liu H, Chen S, et al. Preparation of N-TiO_2/RGO nanocomposites through sol-gel method. Korean Journal of Chemical Engineering. 2021;**38**(9):1913-1922

[29] Bai S, Shen X. Graphene–inorganic nanocomposites. RSC Advances. 2012;**2**(1):64-98. DOI: 10.1039/C1RA00260K

[30] Fan Y, Liu HJ, Zhang Y, Chen Y. Adsorption of anionic MO or cationic MB from MO/MB mixture using polyacrylonitrile fiber hydrothermally treated with hyperbranched polyethylenimine. Journal of Hazardous Materials. 2015;**283**:321-328. Available from: https://www.sciencedirect.com/science/article/pii/S0304389414007870

[31] Hardiansyah A, Budiman WJ, Yudasari N, Isnaeni KT, Wibowo A. Facile and green fabrication of microwave-assisted reduced graphene oxide/titanium dioxide nanocomposites as photocatalysts for rhodamine 6G degradation. ACS Omega. 2021;**6**(47):32166-32177. DOI: 10.1021/acsomega.1c04966

[32] Tsuzuki T. Mechanochemical synthesis of metal oxide nanoparticles. Communications Chemistry. 2021;**4**(1):143. DOI: 10.1038/s42004-021-00582-3

[33] Posudievsky OY, Kondratyuk AS, Kozarenko OA, Cherepanov VV, Dovbeshko GI, Koshechko VG, et al. Facile mechanochemical preparation of nitrogen and fluorine co-doped graphene and its electrocatalytic performance. Carbon N Y. 2019;**152**:274-283. Available from: https://www.sciencedirect.com/science/article/pii/S0008622319305937

[34] González VJ, Vázquez E, Villajos B, Tolosana-Moranchel A, Duran-Valle C, Faraldos M, et al. Eco-friendly mechanochemical synthesis of titania-graphene nanocomposites for pesticide photodegradation. Separation and Purification Technology. 2022;**289**:120638. Available from: https://www.sciencedirect.com/science/article/pii/S1383586622001988

[35] Mishra RK, Mentha SS, Misra Y, Dwivedi N. Emerging pollutants of severe environmental concern in water and wastewater: A comprehensive review on current developments and future research. Water-Energy Nexus. 2023;**6**:74-95. Available from: https://www.sciencedirect.com/science/article/pii/S2588912523000140

[36] Tavengwa NT, Moyo B, Musarurwa H, Dalu T. Chapter 20 - Challenges and future directions in the analysis of emerging pollutants in aqueous environments. In: Dalu T, Tavengwa NT, editors. Emerging Freshwater Pollutants. Netherlands: Elsevier; 2022. pp. 373-379. Available from: https://www.sciencedirect.com/science/article/pii/B9780128228500000168

[37] Chowdhury I, Walker SL, Mylon SE. Aggregate morphology of nano-TiO_2:

Role of primary particle size, solution chemistry, and organic matter. Environmental Science. Processes & Impacts. 2013;**15**(1):275-282. DOI: 10.1039/C2EM30680H

[38] Almquist CB, Biswas P. Role of synthesis method and particle size of nanostructured TiO_2 on its photoactivity. Journal of Catalysis. 2002;**212**(2):145-156. Available from: https://www.sciencedirect.com/science/article/pii/S0021951702937838

[39] Suttiponparnit K, Jiang J, Sahu M, Suvachittanont S, Charinpanitkul T, Biswas P. Role of surface area, primary particle size, and crystal phase on titanium dioxide nanoparticle dispersion properties. Nanoscale Research Letters. 2010;**6**(1):27. Available from:. DOI: 10.1007/s11671-010-9772-1

[40] Yu W, Zhao L, Chen F, Zhang H, Guo LH. Surface bridge hydroxyl-mediated promotion of reactive oxygen species in different particle size TiO_2 suspensions. Journal of Physical Chemistry Letters. 2019;**10**(11):3024-3028. DOI: 10.1021/acs.jpclett.9b00863

[41] Nasikhudin DM, Kusumaatmaja A, Triyana K. Study on photocatalytic properties of TiO_2 nanoparticle in various pH condition. Journal of Physics Conference Series. 2018;**1011**(1):012069. DOI: 10.1088/1742-6596/1011/1/012069

[42] Santhi K, Navaneethan M, Harish S, Ponnusamy S, Muthamizhchelvan C. Synthesis and characterization of TiO_2 nanorods by hydrothermal method with different pH conditions and their photocatalytic activity. Applied Surface Science. 2020;**500**:144058. Available from: https://www.sciencedirect.com/science/article/pii/S0169433219328740

[43] Kocijan M, Ćurković L, Gonçalves G, Podlogar M. The potential of rGO@ TiO_2 photocatalyst for the degradation of organic pollutants in water, Sustainability. Vol. 14. Switzerland, MDPI; 2022

[44] Yadav A, Yadav M, Gupta S, Popat Y, Gangan A, Chakraborty B, et al. Effect of graphene oxide loading on TiO_2: Morphological, optical, interfacial charge dynamics-A combined experimental and theoretical study. Carbon N Y. 2019;**143**:51-62 Available from: https://www.sciencedirect.com/science/article/pii/S0008622318310091

[45] Fu Z, Zhang S, Fu Z. Preparation of multicycle GO/TiO_2 composite photocatalyst and study on degradation of methylene blue synthetic wastewater. Applied Sciences (Switzerland). 2019;**9**(16):3282

[46] Yadav SM, Desai MA, Sartale SD. Superoxide ($O_2^{\bullet-}$) radical species driven type II TiO_2/*g*-C_3N_4 heterojunction photocatalyst for RhB dye degradation. Journal of Materials Science: Materials in Electronics. 2023;**34**(22):1651. DOI: 10.1007/s10854-023-11019-z

[47] Najafi M, Kermanpur A, Rahimipour MR, Najafizadeh A. Effect of TiO_2 morphology on structure of TiO_2-graphene oxide nanocomposite synthesized via a one-step hydrothermal method. Journal of Alloys and Compounds. 2017;**722**:272-277. Available from: https://www.sciencedirect.com/science/article/pii/S0925838817319758

[48] Singh G, Ubhi MK, Jeet K, Singla C, Kaur M. A Review on Impacting Parameters for Photocatalytic Degradation of Organic Effluents by Ferrites and their Nanocomposites, Processes. Vol. 11. Switzerland: Multidisciplinary Digital Publishing Institute (MDPI); 2023

[49] Loeb SK, Alvarez PJJ, Brame JA, Cates EL, Choi W, Crittenden J, et al. The

technology horizon for photocatalytic water treatment: Sunrise or sunset? Environmental Science & Technology. 2019;**53**(6):2937-2947. DOI: 10.1021/acs.est.8b05041

[50] Deshmukh SP, Kale DP, Kar S, Shirsath SR, Bhanvase BA, Saharan VK, et al. Ultrasound assisted preparation of rGO/TiO_2 nanocomposite for effective photocatalytic degradation of methylene blue under sunlight. Nano-Structures & Nano-Objects. 2020;**21**:100407. Available from: https://www.sciencedirect.com/science/article/pii/S2352507X19303300

[51] Zawawi A, Ramli RM, Yub HN. Photodegradation of 1-Butyl-3-methylimidazolium chloride [Bmim] Cl via synergistic effect of adsorption–Photodegradation of Fe-TiO_2/AC. Technologies (Basel). 2017;**5**(4):82

[52] Ahmed SN, Haider W. Heterogeneous photocatalysis and its potential applications in water and wastewater treatment: A review. Nanotechnology. 2018;**29**(34):342001. DOI: 10.1088/1361-6528/aac6ea

[53] Chang J, Zhang Q, Liu Y, Shi Y, Qin Z. Preparation of Fe_3O_4/TiO_2 magnetic photocatalyst for photocatalytic degradation of phenol. Journal of Materials Science: Materials in Electronics. 2018;**29**(10):8258-8266. DOI: 10.1007/s10854-018-8832-7

[54] Thongam DD, Chaturvedi H. Advances in nanomaterials for heterogeneous photocatalysis. Nano Express. 2021;**2**(1):012005. DOI: 10.1088/2632-959X/abeb8d

[55] Isai KA, Shrivastava VS. Photocatalytic degradation of methylene blue using ZnO and 2% Fe-ZnO semiconductor nanomaterials synthesized by sol-gel method: A comparative study. Journal of Water and Environmental Nanotechnology. 2019;**4**(3):251-262

[56] Chiu YH, Chang TFM, Chen CY, Sone M, Hsu YJ. Mechanistic insights into photodegradation of organic dyes using heterostructure photocatalysts. Catalysts. 2019;**9**(5):430

[57] Vrinceanu N, Hlihor RM, Simion AI, Rusu L, Fekete-Kertész I, Barka N, et al. New evidence of the enhanced elimination of a persistent drug used as a lipid absorption inhibitor by advanced oxidation with UV-A and nanosized catalysts. Catalysts. 2019;**9**(9):761

[58] Ateş S, Baran Aydın E, Yazıcı B. Characterization and photocatalytic activity of Al_2O_3 nanopores/$InSnO_2$ electrode for methyl orange degradation. Journal of Materials Science: Materials in Electronics. 2020;**31**(17):14691-14701. DOI: 10.1007/s10854-020-04032-z

[59] Sangareswari M, Meenakshi SM. Development of efficiency improved polymer-modified TiO_2 for the photocatalytic degradation of an organic dye from wastewater environment. Applied Water Science. 2017;**7**(4):1781-1790. DOI: 10.1007/s13201-015-0351-6

[60] Deng Y. Developing a Langmuir-type excitation equilibrium equation to describe the effect of light intensity on the kinetics of the photocatalytic oxidation. Chemical Engineering Journal. 2018;**337**:220-227. Available from: https://www.sciencedirect.com/science/article/pii/S1385894717321800

[61] Zhao W, Adeel M, Zhang P, Zhou P, Huang L, Zhao Y, et al. A critical review on surface-modified nano-catalyst application for the photocatalytic degradation of volatile organic compounds. Environmental Science: Nano. 2022;**9**(1):61-80. DOI: 10.1039/D1EN00955A

[62] Sang Y, Liu H, Umar A. Photocatalysis from UV/Vis to near-infrared light: Towards full solar-light spectrum activity. ChemCatChem. 2015;7(4):559-573. DOI: 10.1002/cctc.201402812

[63] Jo WK, Tayade RJ. New generation energy-efficient light source for photocatalysis: LEDs for environmental applications. Industrial and Engineering Chemistry Research. 2014;**53**(6):2073-2084. Available from:. DOI: 10.1021/ie404176g

[64] Reza KM, Kurny ASW, Gulshan F. Parameters affecting the photocatalytic degradation of dyes using TiO_2: A review. Applied Water Science. 2017;7(4):1569-1578. DOI: 10.1007/s13201-015-0367-y

[65] Ollis DF. Solar-assisted photocatalysis for water purification: Issues, data, questions. In: Pelizzetti E, Schiavello M, editors. Photochemical Conversion and Storage of Solar Energy. Dordrecht: Springer Netherlands; 1991. pp. 593-622

[66] Hussein FH. Photochemical treatments of textile industries wastewater. In: Hauser PJ, editor. Advances in Treating Textile Effluent. Rijeka: IntechOpen; 2011. p. 6. DOI: 10.5772/18902

[67] Pirgholi-Givi G, Farjami-Shayesteh S, Azizian-Kalandaragh Y. The influence of irradiation intensity and stirring rate on the photocatalytic activity of titanium dioxide nanostructures prepared by the microwave-assisted method for photodegradation of MB from water. Physica B: Condensed Matter. 2020;**578**

[68] Kocijan M, Vukšić M, Kurtjak M, Ćurković L, Vengust D, Podlogar M. TiO_2-based heterostructure containing *g*-C_3N_4 for an effective photocatalytic treatment of a textile dye. Catalysts. 2022;**12**(12):1554

[69] Zhang X, Yang Y, Ngo HH, Guo W, Wen H, Wang X, et al. A critical review on challenges and trend of ultrapure water production process. Science of the Total Environment. 2021;**785**:147254. Available from: https://www.sciencedirect.com/science/article/pii/S0048969721023251

[70] Lado Ribeiro AR, Moreira NFF, Li Puma G, Silva AMT. Impact of water matrix on the removal of micropollutants by advanced oxidation technologies. Chemical Engineering Journal. 2019;**363**:155-173. Available from: https://www.sciencedirect.com/science/article/pii/S1385894719300968

[71] Heredia Deba SA, Wols BA, Yntema DR, Lammertink RGH. Effects of the water matrix on the degradation of micropollutants by a photocatalytic ceramic membrane. Membranes (Basel). 2022;**12**(10):1004

[72] Kocijan M, Ćurković L, Vengust D, Radošević T, Shvalya V, Gonçalves G, et al. Synergistic remediation of organic dye by titanium dioxide/reduced graphene oxide nanocomposite. Molecules. 2023;**28**(21):7326

[73] Kusiak-Nejman E, Morawski AW. TiO_2/graphene-based nanocomposites for water treatment: A brief overview of charge carrier transfer, antimicrobial and photocatalytic performance. Applied Catalysis. B, Environmental. 2019;**253**:179-186. Available from: https://www.sciencedirect.com/science/article/pii/S0926337319303777

[74] Ch-Th T, Manisekaran R, Santoyo-Salazar J, Schoefs B, Velumani S, Castaneda H, et al. Graphene oxide decorated TiO_2 and $BiVO_4$ nanocatalysts for enhanced visible-light-driven

photocatalytic bacterial inactivation. Journal of Photochemistry and Photobiology A: Chemistry. 2021;**418**:113374. Available from: https://www.sciencedirect.com/science/article/pii/S101060302100246X

[75] Zhou X, Zhou M, Ye S, Xu Y, Zhou S, Cai Q, et al. Antibacterial activity and mechanism of the graphene oxide (rGO)-modified TiO_2 catalyst against Enterobacter hormaechei. International Biodeterioration & Biodegradation. 2021;**162**:105260. Available from: https://www.sciencedirect.com/science/article/pii/S0964830521000901

[76] Wanag A, Rokicka P, Kusiak-Nejman E, Kapica-Kozar J, Wrobel RJ, Markowska-Szczupak A, et al. Antibacterial properties of TiO_2 modified with reduced graphene oxide. Ecotoxicology and Environmental Safety. 2018;**147**:788-793. Available from: https://www.sciencedirect.com/science/article/pii/S0147651317306139

Chapter 7

Perspective Chapter: TiO_2 Electron Transporting Layers for Perovskite Solar Cells

Abimbola Jacob Olasoji and Sang Hyuk Im

Abstract

TiO_2 is a very useful material for the fabrication of solar cells such as dye-sensitized solar cells, quantum dot-sensitized solar cells, organic-inorganic hybrid solar cells, and perovskite solar cells. Among these, perovskite solar cells have been of great interest over the last decade because of their prominent properties such as high absorptivity, ambipolar charge transportability, convenient bandgap tunability, and solution processability. To obtain high-performance perovskite solar cells (PSCs), using effective electron transport layers (ETLs) of TiO_2 is crucial to ensure efficient charge separation, which occurs mainly at the interface between the ETL structure and the perovskite photoactive layer. Therefore, this chapter will introduce TiO_2 ETLs and cover how to prepare and modify the TiO_2 ETLs to achieve high-efficiency perovskite solar cells.

Keywords: titanium oxide, electron transporting layers, TiO_2 synthesis, TiO_2 deposition, perovskite solar cells

1. Introduction

Metal halide perovskites (MHPs) have been of great interest over the last decade because of their excellent properties, such as high absorptivity due to direct bandgap, convenient bandgap tuneability by compositional and structural engineering, long carriers' lifetime by low trap density, high open circuit voltage due to the formation of shallow traps, and solution processability. Metal halide perovskite solar cells are generally composed of transparent conducting oxide (TCO) substrate/electron transport layer (ETL)/MHP/hole transport layer (HTL)/counter electrode. These perovskite solar cells (PSCs) can be classified, for example, into mesoscopic, bi-layer, and planar types in terms of device architecture, as shown in **Figure 1** [1–4]. The basic operating mechanisms of these mesoscopic, bi-layer, and planar type PSCs are the same because the electrons and holes are generated in the perovskite layer, and the electrons are transported into TCO through both the ETL and perovskite layer, while holes are transported into the counter electrode through HTL. Hence, irrespective of the ETL form used, the electrons generated in the perovskite layer are transported into ETL. However, the structure of the ETL can be a contributing factor to improving transportation. Time-resolved photoluminescence (TRPL) spectra of $MAPbI_3$ (methylammonium lead triiodide) perovskite, perovskite/TiO_2 ETL, perovskite/PTAA (poly-triarylamine) HTL, and ETL/perovskite/HTL samples

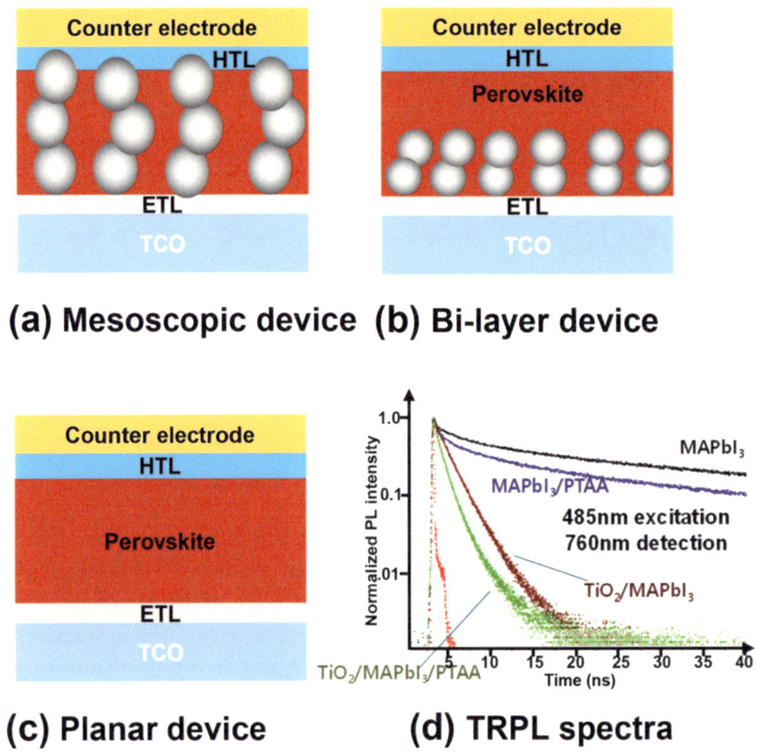

Figure 1.
(a–c) Schematic illustration of device architecture of (a) mesoscopic, (b) bi-layer, and (c) planar PSCs and (d) time-resolved photoluminescence (TRPL) spectra of perovskite, perovskite/ETL, perovskite/HTL, and ETL/perovskite/HTL samples (ETL = TiO_2, HTL = PTAA (poly-triarylamine)).

in **Figure 1(d)** confirm that the greatest photoluminescence (PL) quenching happens at the TiO_2 ETL/perovskite interface [5]. This indicates that when illuminated light enters through the TCO/ETL side, charge separation occurs mainly at the TiO_2 ETL/perovskite interface. Hence, physical as well as chemical and electronic properties of ETL are very important to determine and enhance the performance of PSCs.

Liquid junction mesoscopic perovskite-sensitized solar cells were first reported by the Miyasaka group in 2009. The mesoscopic devices were composed of mesoscopic TiO_2 ETL/$MAPbI_3$ or $MAPbBr_3$ perovskite/liquid electrolyte/counter electrode, and their power conversion efficiencies (PCEs) were 3.8 and 3.1%, respectively [6]. However, their device stabilities were very poor because perovskite is quickly corroded by liquid electrolyte. To solve this problem, a solid HTL such as spiro-Meeta (2,2′,7,7′-tetrakis(*N*,*N*-di-*p*-methoxyphenylamine)-9,9′-spirobifluorene) was used in the mesoscopic PSCs, and a remarkable 9.7% of PCE with improved stability was achieved [7]. Heo et al. reported that mesoscopic TiO_2 PSCs are heterojunction solar cells with main charge separation taking place at mesoscopic TiO_2/$MAPbI_3$ perovskite junction [8]. As coating technology for the formation of pinhole-free perovskite film emerged, mesoscopic device configuration was changed to bi-layer or planar type. Currently, the PCEs of single-junction PSCs and PSC-Si tandem solar cells have reached 26.1 and 33.9%, respectively [9].

TiO_2 is the most widely used ETL material by different research groups for PSCs. An ETL serves the purpose of transferring electrons and blocking the transport of holes to the TCO electrode in PSCs. An eligible ETL must have an energy band position that aligns with the perovskite photoactive layer such that it is upper conduction band position should be slightly below the conduction band minimum of the perovskite layer to ensure electron injection, while the lower valence band stands at a deeper position to efficiently block holes. Also, the ETL should have high transparency for strong light transmittance and uniform film morphology to reduce losses due to leakage. An ideal ETL promotes high-quality perovskite film by ensuring well-crystalized pinhole-free compact morphology, effective electron transfer, and lowers recombination tendencies and charge losses, as it indirectly curbs the hysteresis effect and enhances device stability [10, 11]. Metal oxides that are often used as ETL include but are not limited to TiO_2, ZnO, SnO_2, Nb_2O_5, WO_x, and WO_x-TiO_x [12–19]. Among these, TiO_2 is the most commonly used ETL in PSCs due to its suitable band structure, good electron mobility, non-toxicity, stability in air, and low cost [20–23].

TiO_2 is an n-type metal oxide semi-conductor which has wide band gap of ~3 eV. Polymorphs of TiO_2 exist as rutile, anatase, and brookite in their natural and synthetic crystalline forms, with anatase and rutile being the thermodynamically stable crystal phases. Like other transition metal oxides, oxygen vacancies in TiO_2 serve as sources of structural point defects where oxygen atoms are missing within the structure as interstitial atoms fill the cavity space between the regular atomic sites. Vacancy is formed by transferring oxygen atoms from atomic site to gaseous state. The oxygen vacancy point defects foster electrical conductivity by acting as electron donors, aiding ionization, and also generating ionic current in response to an electric field [24].

2. Deposition processes for the preparation of TiO_2 ETL

Following an increase in demand for different types of TiO_2 ETLs in perovskite solar cells, a range of synthesis methods for the ETL have emerged over the years, as shown in **Figure 2**, that can be used to enhance electron transporting and light harvesting efficiency in perovskite solar cells. Generally, methods of preparing TiO_2 are broadly classified under solution chemistry and vapor deposition methods.

2.1 Solution phase depositions

Chemical strategies that utilize solution chemistry techniques, which are otherwise known as wet processes, include sol-gel chemistry, co-precipitation, hydrothermal, etc. The standard mechanisms for some examples of synthesis approaches used for fabrication of semi-conductor metal oxide nanomaterials that provide excellent control over many of the governing parameters are shown in **Figure 3**.

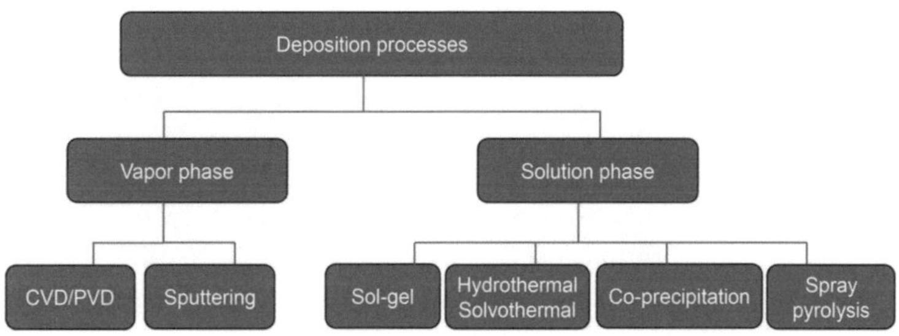

Figure 2.
Deposition processes used for preparation of TiO_2 ETL.

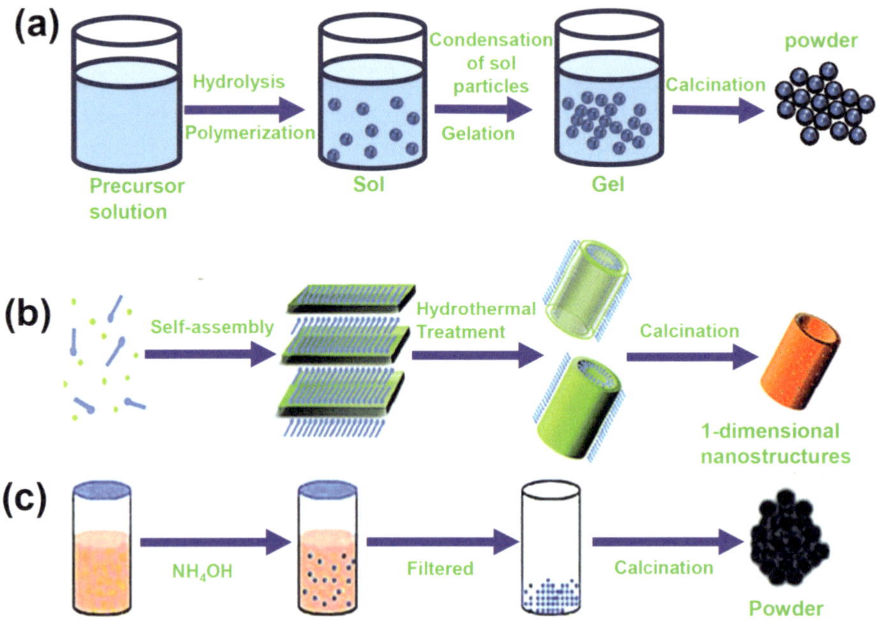

Figure 3.
Schematic representation of the chemical synthesis routes of semi-conductor metal oxide nanomaterials via (a) sol-gel, (b) hydrothermal, and (c) co-precipitation methods. Reproduced from Ref. [25] with permission from the Royal Society of Chemistry licensed under a creative commons attribution.

2.1.1 Sol-gel method

The sol-gel method is commonly used to synthesize metal oxides such as TiO_2 nanostructured materials, which have excellent optical and electronic properties useful for photovoltaic energy conversion. This strategy consists of hydrolysis and polymerization of an inorganic metal salt or metal organic compound known as a precursor (typically metal alkoxides) to form an amorphous polymeric network of titanium oxides and hydroxides upon completion of the polymerization step. The liquid gel obtained can further undergo hydrothermal treatment or be used as it is to produce TiO_2 thin films (by spin-coating, dip-coating, or chemical bath deposition) and 3D porous scaffolds (by using 3D templates). The presence of water in the colloidal suspension/precursor solution of $Ti(OR)_x$ leads to hydrolysis of the precursor. Building of Ti-O-Ti bonds and further polymerization of the precursor can be promoted by the addition of catalytic agents (usually solutions of inorganic acid) to enhance peptization of the colloid. Adjustment of experimental parameters such as pH, temperature, and water content makes it possible to achieve the formation of long polymeric Ti-O-Ti chains that possess a narrow size distribution [26].

The three major techniques of depositing TiO_2 on substrates using precursor solutions include spin-coating, dip-coating, and chemical bath deposition (CBD). Spin-coating involves rapid deposition of thin film onto a substrate held by some rotatable spindle where vacuum is used to fix the substrate into place, and the coating fluid is centrifugally spread on the surface radially outwards based on spin speed applied until the thin film is formed. While the initial volume of dispensed fluid on the substrate and dispensation rate have very little effect on the final thickness of the film, the thickness is controlled by key parameters such as spin speed (rotation velocity) and viscosity of the solution. Dip-coating is done by deposition of a wet liquid film through gradual withdrawal of a substrate from the solution medium. This method is economically advantageous and simple to do. Key parameters for this deposition are the viscosity and withdrawal speed of the substrate during deposition [24]. The chemical bath deposition technique has advantages such as cheap cost, large-area adaptability, and no need for complex instruments. Factors that affect film growth rate, film thickness, and quality of crystallites on substrate include pH of solution, immersion time, temperature, and bath concentration [27].

In 2016, Liang et al. used a chemical bath to deposit a rutile TiO_2 compact layer on the substrate to achieve an efficient planar heterojunction perovskite solar cell (with a PCE of 12.62%). The rutile TiO_2 thin films were grown on fluorine-doped tin oxide (FTO) substrate at 70°C using an optimized concentration of 200 mM $TiCl_4$ precursor solution and subsequent UV/O_3 treatment to reduce carrier charge recombination to achieve over 12% PCE. This work offers use in flexible substrate perovskite solar cells where low temperature is required to deposit TiO_2 thin film as an effective transport layer [28].

Ma et al. applied sol-gel chemistry to fabricate craterlike dual-functioning (porous/blocking bi-layer) TiO_2 ETL for high-efficiency perovskite solar cells (with a PCE of 16%) in 2018. Ti alkoxide-based sol-gel ink consisting of titanium isoprenoid ($Ti(OPr)_4$) modified with acetylacetone (Acacia), ammonium nitrate (AN), and 2-methoxyethanol (2ME) was used in the spin-coating deposition process of TiO_2 films, which have craterlike surface pores of 100–300 nm diameter and ~50 nm depth. The presence of the craterlike porous structure decreased the difference between refractive indices of fluorine-doped tin oxide (FTO) and TiO_2, thus leading to a reduction in reflectance and increasing light transmission. Interstitially doped nitrogen was also noticed in the structure lattice, which increased the conductivity of the ETL. Compared to conventional dense TiO_2 based PSC, this craterlike porous/blocking bi-layer-structured ETL showed an improvement of 14.5% in photocurrent density and 19.5% enhancement in PCE [29].

2.1.2 Hydrothermal and solvothermal methods

Processes that involve hydrothermal and solvothermal processes rely on the use of an autoclave at high pressure and moderate temperature. A process is said to be hydrothermal or solvothermal if the solvent used is aqueous or organic respectively. A unique characteristic of both techniques is the elevated range of processing temperature, which operates above the boiling point of the solvent. Also, the crystallinity of particles synthesized through the sol-gel route can be improved through hydrothermal and solvothermal methods. For instance, it was discovered that 1 hour hydrothermal processing at 250°C drastically improved the crystallinity of the TiO_2 nanoparticles that were previously prepared via three different sol–gels using TTIP or $TiCl_4$ as precursors [30]. It was also reported that at 240°C in an autoclave, well-dispersed and phase-pure anatase TiO_2 nanoplatelets via hydrolysis of $TiCl_4$ solution were synthesized using ethylene glycol solvent via solvothermal method [31].

Liu et al. prepared modified anatase TiO_2 nanoparticles (NPs) via hydrothermal method at 150°C for low-temperature fabrication of efficient ETL in perovskite solar cells, and the cells exhibited reduced charge transport resistance, enhanced electron extraction, and suppressed charge recombination. The anatase TiO_2 NPs of 15 nm average diameter were synthesized using a tetrabutyl titanate precursor in an ethanol solvent and subsequently modified using acetic acid (AA) as well as oleic acid (OA).

The modified NPs dispersed in precursor solution were spin-coated into TiO_2 ETL thin film to make PSC of 20.15 and 19.41% PCE with AA and OA, respectively [32].

Supraja et al. synthesized biphasic TiO_2 NPs using the solvothermal method for the fabrication of TiO_2 ETL to enhance the photovoltaic performance and stability of perovskite solar cells. The NPs in the biphasic TiO_2 mixture were synthesized in an autoclave at 180°C using titanium precursor in n-propanol solvent, hydrolyzed with deionized (DI) water, and catalyzed with HCl. An optimum of 75% brookite and 25% rutile TiO_2 particles solution spin-coated into TiO_2 ETL in PSC exhibited the highest PCE of 14% which has an open circuit voltage of 1.1 V and a current density of 17.4 mA/cm^2 [33].

2.1.3 Co-precipitation method

Co-precipitation is a chemical strategy technique for fabricating TiO_2 nanostructures such as nanoparticles and nanorods based on the condition of using a supersaturated aqueous solution to precipitate metal hydroxides and subsequent thermal treatment to crystallize the oxide. It is a wet chemical solution method that is scale-up adaptable. The major factors that affect the synthesis of micro- or nanostructures of multiphase TiO_2 include reaction conditions, reactivity of starting or intermediate materials, pH of the reaction medium, and reaction parameters such as temperature and time of reaction [25].

Maryam et al. employed the use of co-precipitation to synthesize TiO_2 nanoparticles used in the TiO_2/ZnO photoanode as ETL material in a perovskite solar cell. The ETL film was achieved by the screen-printing method using composited TiO_2 and ZnO based on variations of the TiO_2/ZnO powder mass ratio to combine the advantages of both materials. The TiO_2 powder was synthesized by reacting $TiCl_3$, HCl, water, and ammonia for 24 hours until TiO_2 residue is precipitated, washed, dried at 100°C, and annealed at 450°C for 2 hours. The PSC with higher TiO_2 had good and stable photosensitivity [34].

2.1.4 Spray pyrolysis deposition method, SPD

While gas phase methods employ a chemical vapor deposition process, the spray pyrolysis method directly employs the use of TiO_2 precursor solutions by spraying the metal salt solution onto a heated substrate. In this case, the surface temperature of the substrate is the key parameter, as it influences the film roughness, cracking, and crystallinity of the thin film deposited. As droplets fall on the heated substrate surface, they usually spread into a disk-shaped structure as they undergo thermal decomposition. Spray pyrolysis deposition is a cost-effective and easy method employed for the preparation of dense films. Major components of spray pyrolysis equipment include a substrate heater, a precursor solution-fed atomizer (hand-spray or ultrasonic), and a temperature controller [35].

Sun et al. utilized ultrasonic spray deposition of TiO_2 ETL as a hole-blocking layer to fabricate reproducible and high-efficiency hybrid perovskite solar cells. Unlike the simple airbrush approach, scalable ultrasonic spray was adopted to ensure film homogeneity across larger device areas. The dense TiO_2 blocking layer was fabricated on FTO substrate heated at 500°C by ultrasonic spray pyrolysis of titanium (IV) diisopropoxide bis(acetylacetonate) ($Ti(acac)_2OiPr_2$) precursor solution for about 10 minutes, after which it was let to cool down. Compared to hand-spray deposition of TiO_2 with macroscopic mean thickness deviation of up to ±15%, ultrasonic spray yielded coatings of significantly reduced thickness variation to less than ±5%. Coupling the ultrasonic spray (200 kPa) blocking layer to a mesoporous TiO_2 nanoparticle scaffold yielded improved charge transport with a reproducible PCE of 17.4% [36].

2.2 Vapor phase depositions

In the vapor deposition method, materials in their gaseous state are deposited as functional materials to yield excellent thin film nano/microstructure through chemical reaction or condensation. Also, this deposition requires controllable transfer of atoms for film growth to proceed epitaxially. Vapor deposition techniques (also known as dry processes) include chemical vapor deposition, physical vapor deposition, sputter deposition, etc. These techniques are used to deposit uniformly thin films. However, they are quite energy intensive and release poisonous gas as their by-products [37].

2.2.1 Chemical, atomic layer, and physical vapor deposition methods (CVD, ALD, PVD)

Atomic layer deposition (ALD) is a type of chemical vapor deposition (CVD) in which thin films are fabricated by stacking atomic layers one after the other. CVD and PVD deposition methods employ the use of high temperatures under vacuum to vaporize liquid or solid precursors and subsequently condense the resulting vapor into solid-phase material on a substrate. Condensation takes place due to the difference in temperature between the plasma used to vaporize the precursor and the substrate. These methods have been used to successfully deposit nanostructured TiO_2 thin films on different substrates. However, unlike CVD, no chemical reaction takes place during deposition via PVD. A case study of well-aligned and perpendicularly oriented homogenous TiO_2 nanorods (200–300 nm) thin films was achieved by condensation of a titanium acetylacetonate precursor vaporized at 500–700°C in an N_2/O_2 gas flow on a silica substrate [38]. Though both methods yielded good results for TiO_2 deposition, high costs of production due to temperature conditions and expensive equipment constitute significant drawbacks for their use. ALD is characterized by better film uniformity and thickness controllability. Hence, the ALD method is more frequently used (among the other vapor deposition methods) for the fabrication of TiO_2 ETL used in perovskite solar cells.

Shalan et al. fabricated a uniform and pinhole-free blocking layer of TiO_2 via atomic layer deposition for mesoporous perovskite solar cells. The TiO_2 blocking layer deposition was done inside an ALD reactor by alternating exposure of $TiCl_4$ and deionized water vapor at a process temperature of 300°C at 1.6 kPa for a precursor carrier and N_2 purge gas operating at pulse and purge times of 0.1 and 4 seconds, respectively, while giving an atomic deposition rate of 0.43 Å per cycle. Mesoporous m-TiO_2 was further spin-coated and annealed at 450°C for 30 minutes. At an optimized thickness of 200 nm, the ALD-TiO_2 compact layer promoted efficient electron transfer with the highest PCE of 15% [39].

2.2.2 Sputtering method

Production of a TiO_2 compact layer can also be done via Ti sputter deposition. This is achieved via radio frequency (RF) reactive magnetron sputtering from a high-purity Ti target in an $Ar/O_2/N_2$ atmosphere, which is deposited on a substrate following subsequent oxidation. A case study of deposited compact TiO_2 involves using constant sputtering pressure (10 mTorr), RF power (100 W), substrate temperature (300°C), variable flow rates for argon, oxygen, and nitrogen, and an 80 mm distance between the substrate and the target [40]. To achieve the Ti sputtering process, there is a need for very expensive equipment and the provision of a strong vacuum. Hence, this method is not readily available for use [41].

Ke et al. applied TiO_2 compact film fabricated by thermal oxidation of sputtered Ti film. The Ti films were coated on FTO glasses by radio frequency (RF) magnetron sputtering from a Ti target (99.9% purity) with RF power and sputtering pressure at 100 W and 1 Pa, respectively. Thermal oxidation into the TiO_2 compact layer took place during the one-step sintering process together with the spin-coated TiO_2 porous layer at 500°C for 30 minutes. At an optimized thickness of 15 nm of dense compact layer, the recombination process at the interface in PSC was greatly inhibited to give the highest PCE of 15.07% [42].

3. Types of TiO_2 ETL nanostructures

Various types of TiO_2 nanostructures have also attracted much attention for application in perovskite solar cells. Jarwal et al. produced TiO_2 nanorod arrays with improved surface-to-volume ratio and direct carrier transportation [43] by solvothermal etching and/or $TiCl_4$ treatment. Nanowire arrays and nanoflower composites of TiO_2 film were fabricated by Lu et al. [44] using $TiCl_4$ precursor solution. The resulting composite structure showed improved light harvest, short direct charge transmission path, small leakage current and charge transmission resistance, and slow recombination rate. Also, doping is an excellent method for improving the quality of TiO_2 films. The commonly doped materials [45] include but are not limited to Nb [46], Ta [47], Li [48], Ce [49], Mg [50], Zn [51], $EuAc_3$ [52], Zr [53], and graphene quantum dots (QDs) [54]. Bi-layer ETLs consisting of TiO_2 are also employed, like the following: TiO_2/NiO [55], TiO_2/ZnO [56], TiO_2/SnO_2 [57], TiO_2/graphene [58], TiO_2/WO_3 [59], and TiO_2/fullerene [60].

3.1 TiO_2 nanorods, nanotubes and nanowires

According to recent reports, tuning of size and shape of TiO_2 nanostructures can be achieved under high degree of synthetic control from spherical nanoparticles to nanostructures such as 1-dimensional (1D) nanorods, nanowire arrays, and nanotubes. Bottom-up routes of fabricating enormously complex nanostructures have attracted key interest due to their versatility and potential to fabricate good-quality TiO_2 nanostructures. Routes such as sol-gel, hydrothermal, and co-precipitation methods achieve excellent chemical synthesis of multiphase TiO_2 nanostructures due to (1) good control of structural phases and stoichiometry, (2) narrow dispersion in homogeneity, size, and morphology, and (3) large-scale adaptability, inexpensiveness, time efficiency and low impurity in the production of nanomaterials [25].

Figure 4a [61] shows the Scanning electron microscopy (SEM) morphology of anatase-TiO_2 nanorods by the co-precipitation method [25]. TiO_2 nanowires were synthesized via the hydrothermal method, as shown by the SEM micrograph in **Figure 4b** [61]. Also, SEM and transmission electron microscopy (TEM) images of TiO_2 nanotubes synthesized through the hydrothermal method are shown in **Figure 3c**, which indicates a fibrous texture of length up to 1–3 μm, outer diameters

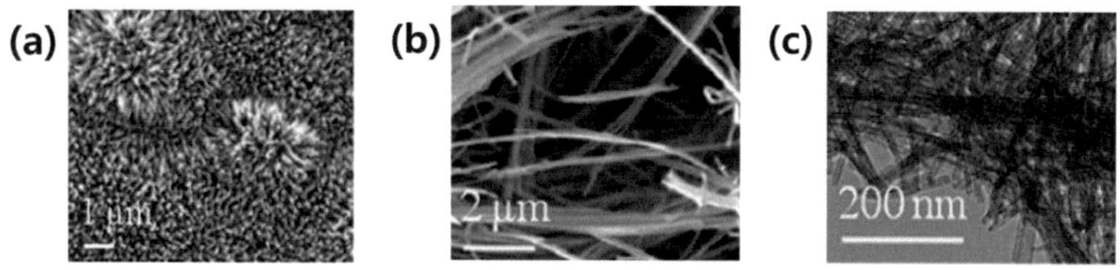

Figure 4.
(a) TiO_2 nanorods [61]: (b) TiO_2 nanowires [61]: (c) TiO_2 nanotubes [62]. Reproduced from Ref. [25] with permission from the Royal Society of Chemistry licensed under a creative commons attribution.

extending between 7 nm and 10 nm, inner diameters at 5 nm, and an inter-wall spacing up to 0.5 nm [62].

Park et al. synthesized novel TiO_2 nanorods (NRs), which serve as excellent building blocks for ETLs of mesoscopic perovskite solar cells. Various lengths of 70–200 nm and uniform widths of 46–48 nm were synthesized via solvothermal reaction in basic medium, and these lengths can be reproducibly tuned by varying concentrations of tetramethylammonium hydroxide (TMAH). Compared to conventional TiO_2 nanoparticles, NRs showed superior photovoltaic performance with fewer defects and significantly reduced defect-mediated charge recombination, as TiO_2 NRs of 110 nm length showed the highest efficiency of 23.18% PCE [63].

Elseman et al. synthesized TiO_2 nanotubes via hydrothermal method as advanced ETL for enhancing the efficiency and stability of perovskite solar cells. TiO_2 NPs of 90 nm were first synthesized by co-precipitation of $Ti(OCH(CH_3)_2)_4$ and then moved to an autoclave for hydrothermal processing at 160°C for 16 hours in a basic medium to produce the TiO_2 nanotubes. After purification, the nanotubes were spin-coated on FTO to make TiO_2 ETL films. Compared to ETLs produced from TiO_2 nanoparticles (PCE of 17.18%), TiO_2 nanotubes suppressed charge recombination and reinforced transport pathways for carriers in PSC to give a higher PCE of 19.14% [64].

Jiang et al. reported a rutile TiO_2 nanowire (NW)-based perovskite solar cell. Single crystal rutile TiO_2 NW arrays were grown on FTO substrate using a tetrabutyl titanate precursor via a ketone-HCl solvothermal process in an autoclave. As a result of high electron diffusivity in the rutile nanowires, electron transport via the TiO_2 ETL was enhanced. At 900 nm long rutile nanowires, which allow additional loading of perovskite material for high light absorption, a higher current density was attained with a PCE of 11.7% which outperforms a counterpart mesoporous-based PSC [65]. Asadzade, Amir, and Shabnam Andalibi Miandoab used FDTD (finite-difference time domain) and FE (finite element) simulation methods to analyze the optical and electrical properties of TiO_2 nanowire scaffold ETL in perovskite solar cells. At a period and height of 100 nm and 400 nm of the nanowires, maximum PCE of 18.9% was achieved [66].

3.2 Porous TiO_2 scaffold (mesoporous TiO_2 and inverse opal TiO_2 ETLs)

Several concerted efforts have been put in place to investigate new strategies to maximize the use of solar energy since solar light is one of the most readily available renewable energy resources on earth. Porous scaffold is considered an option since it could enhance light-matter interactions due to the high surface area inherent in its structure. While mesoporous TiO_2 and inverse opal TiO_2 ETLs are the major examples of porous TiO_2 scaffolds, inverse opal thin films are considered more versatile and worthy of study due to their unique periodic structure that effectively extends light propagation pathways, as they show enormous potential in harvesting solar energy [67].

Inverse opal thin films are photon-manipulation nanostructures, which are a subclass of photonic crystals that have a refractive index differing periodically in space and are characterized by a stopband that prevents transmission of photons within a certain energy range known as a band gap. The high porosity and periodic interconnected pore network structure induces multiple light scattering effects, which promote light flow extension and absorption efficiency of light absorbers deposited in it. This enhanced light-trapping ability, coupled with a high surface area, has the ability to improve the performance of solar cells [68]. The fabrication process of TiO_2 inverse opal follows a bottom-up approach [69], and it consists of major three-steps, which include fabrication of colloidal template by self-assembly of monodisperse nanoparticles of polymers such as polystyrene (PS) or polymethylmethacrylate (PMMA); defect-free infiltration [70] of interstices of the sacrificial template via soaking or spin-assisted infiltration using TiO_2 precursor solution, which solidifies as matrix phase in the composite infiltrated opal template; inversion via etching or calcination, which leaves a replica inverted structure of the colloidal crystal template. The template-directed fabrication method has the tuning ability to vary dimensions of the pores of the inverted thin film by using different sizes of polymer nanospheres in the sacrificial templates [71]. Olasoji et al. from Im group applied template directed process through immersion infiltration (soaking) of opal template in prehydrolyzed $TiCl_4$ precursor solution to prepare crack-free, uniform, open top surface and improved matrix phase quality TiO_2 inverse opal thin-film [70]. A mesoscopic inverse opal (meso-IO) film with a three-dimensionally interconnected porous structure was fabricated and used as an electron-conducting scaffold in perovskite solar cells (**Figure 5**). It was recorded that applying this inverse opal scaffold increased light harvesting efficiency to give an efficiency of 17% PCE together with reduced pinholes in the deposited perovskite layer [72].

3.3 Doped TiO_2

Doping a TiO_2 photoelectrode can help to effectively absorb photons and better enhance band alignment between layers and subsequent electron injection into the

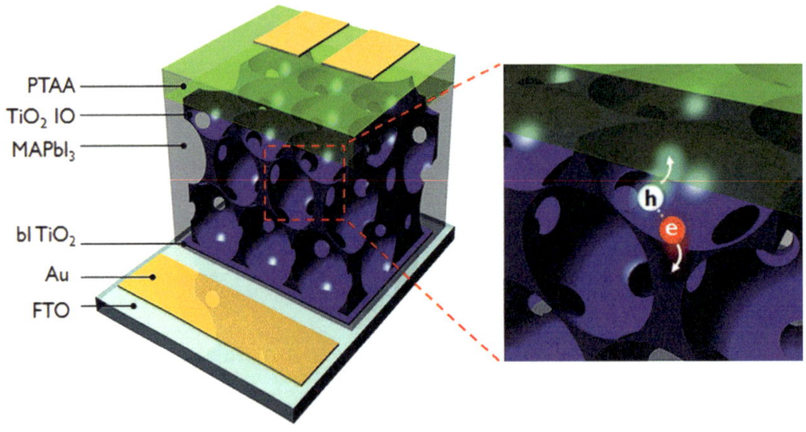

Figure 5.
Schematic of the architecture of the mesoscopic $MAPbI_3$ PSC with TiO_2 inverse opal (IO) electron-conducting scaffold. Reproduced from Ref. [72] with permission from the Royal Society of Chemistry.

electrode. For instance, N-type doping helps to reduce charge recombination by increasing the conduction band and CB edge to allow smooth electron charge injection from the perovskite layer to the ETL [73]. An example is using Y-doped TiO_2 (0.5% Y) to increase the conduction band edge close to the conduction band of perovskite [74]. Also, pure anatase TiO_2 mesoporous ETL has a 3.2 eV band gap with 3.2 eV HOMO and 5.4 eV LUMO. n-doping this TiO_2 ETL improved electron injection from perovskite to the ETL, which resulted in increased V_{OC}, which consequently improved the PCE. With this concept, many n-type doped TiO_2 have been used to achieve higher V_{OC}. n-doping can be done with a few elements, such as Sn, Mg, etc. Zhang et al. [75] doped TiO_2 using Sn to tune up the band gap of TiO_2. This improved the efficiency of the perovskite solar cell by 67%. Also, Manseki et al. doped TiO_2 with Mg to increase the band gap of TiO_2 and CB edge, leading to an increased bandgap of 1 eV and improved V_{oc} from 587 to 802 meV [76]. Giordano et al. n-doped mesoporous TiO_2 ETL via posttreatment with lithium salts to achieve superior electronic properties by reducing electronic trap states, which promoted faster electron transport. Compared to undoped electrodes, the Li-doped TiO_2 ETL films improved the PCE of $CH_3NH_3PbI_3$ perovskite solar cells from 17 to 19% with negligible hysteresis (lower than 0.3%) [77].

3.4 Low deposition temperature TiO_2 ETL for large-area application

Conventional TiO_2 ETL fabrication methods limit large-scale application on flexible substrates due to the high heat treatment (> 450°C) required for the annealing step, which also results in high fabrication costs based on energy consumption. To address this issue, Yoo et al. applied low-temperature UV treatment in place of thermal annealing to sinter thin TiO_2 film prepared via the sputtering deposition method. The resulting sintered TiO_2 ETL showcased enhancements in carrier concentration, electron mobility, hole-blocking ability, defect diminution, and good band alignment compared to thermally sintered films. These sputtered and UV-treated TiO_2 films were applied to large-area perovskite solar cell modules using rigid and flexible substrates, and power conversion efficiencies of 18.82 and 14.61% respectively, were recorded [78].

3.5 High-efficiency TiO_2 ETL-based PSCs

In early-stage research of developing high-efficiency PSCs, TiO_2 ETLs were applied using the normal (n-i-p) configuration to obtain the highest efficiency of over 25% as summarized in **Table 1**, due to effective charge separation at the TiO_2 ETL/perovskite interface. Yang et al. prepared a compact TiO_2 layer via chemical bath deposition to

TiO_2 ETL	Synthesis method	Deposition method	V_{oc} (V)	J_{sc} (mA/cm^2)	FF (%)	PCE (%)
Compact [79]	Sol-gel	CBD	1.18	26.04	82.2	25.3
Mesoporous [80]	Sol-gel	Spin-coating	1.19	26.35	81.7	25.6
Nanorods [63]	Solvothermal	Spin-coating	1.14	24.06	81.0	23.2
Inverse opal [72]	Sol-gel	Template-directed	1.07	20.30	78.8	17.1
Li-doped TiO_2 [77]	Calcination	Doping	1.11	23.00	74.0	19.3

Table 1.
Performance summary of high-efficiency TiO_2 ETL-based PSCs.

fabricate a PSC device and achieved a PCE of 25.3% (certified 24.8%) [79]. Jeong et al. applied mesoporous TiO_2 ETL to PSC and recorded a certified efficiency of 25.2% [80]. Also, Park et al. synthesized 110 nm long TiO_2 nanorods (NRs) ETL for perovskite solar cells to give a PCE of 23.18% [63]. Ha et al. fabricated mesoporous inverse opal scaffold using a template-directed process and applied it to perovskite solar cells at a PCE of 17.1% [72]. Giordano et al. did a lithium doped mesoporous TiO_2 layer in a PSC cell, and they recorded 19.3% efficiency [77]. Lately, TiO_2 ETLs are being gradually replaced with SnO_2 ETLs to fabricate better-performing perovskite solar cells with over 26% PCE because SnO_2 has better transmittance and electron mobility than TiO_2.

4. Conclusion

This chapter provides insights to showcase recent trends on TiO_2 nanostructures, ranging from their synthesis methods to state-of-the-art modification approaches through various chemical routes. Also, the importance and advances in adopting TiO_2 ETL and how the photoactive metal oxide material contributes to the overall enhancement of PSCs for improved efficiency, reduced hysteresis, and increased stability via enhanced charge transfer, reduced defect density, and modified energy (bandgap) levels were explored. Ultimately, this work shows the significance of employing TiO_2 ETL in PSCs. Although TiO_2 has good chemical stability and proper electronic properties for ETL in PSCs, currently it is being gradually replaced with other metal oxide ETLs such as SnO_2, Zn_2SnO_4, etc., because intrinsic TiO_2 has relatively lower conductivity. To expand the utilization of TiO_2 ETL for PSCs, it is necessary to enhance its mobility and conductivity by doping or multiple stacking with other types of ETL. In addition, presently PSC-based multi-junction solar cells have captured great attention to realize highly efficient and cost-effective tandem solar cells with PCE > 35% because they can be used for green hydrogen production, carbon capture, and solar fuel conversion. Following these new frontiers, newly designed and modified TiO_2 ETLs, which can be prepared via solution phase and/or vapor phase depositions, can be utilized.

Author details

Abimbola Jacob Olasoji and Sang Hyuk Im*
Korea University, Seoul, Republic of Korea

*Address all correspondence to: imromy@korea.ac.kr

References

[1] Kim HS, Im SH, Park NG. Organolead halide perovskite: New horizons in solar cell research. The Journal of Physical Chemistry C. 2014;**118**(11):5615-5625

[2] Song DH, Jang MH, Lee MH, Heo JH, Park JK, Sung SJ, et al. A discussion on the origin and solutions of hysteresis in perovskite hybrid solar cells. Journal of Physics D: Applied Physics. 2016;**49**(47):473001

[3] Heo JH, Song DH, Patil BR, Im SH. Recent progress of innovative perovskite hybrid solar cells. Israel Journal of Chemistry. 2015;**55**(9):966-977

[4] Park JK, Heo JH, Han HJ, Lee MH, Song DH, You MS, et al. Efficient hysteresis-less bilayer type $CH_3NH_3PbI_3$ perovskite hybrid solar cells. Nanotechnology. 2015;**27**(2):024004

[5] Heo JH, You MS, Chang MH, Yin W, Ahn TK, Lee SJ, et al. Hysteresis-less mesoscopic $CH_3NH_3PbI_3$ perovskite hybrid solar cells by introduction of Li-treated TiO_2 electrode. Nano Energy. 2015;**15**:530-539

[6] Kojima A, Teshima K, Shirai Y, Miyasaka T. Organometal halide perovskites as visible-light sensitizers for photovoltaic cells. Journal of the American Chemical Society. 2009;**131**(17):6050-6051

[7] Kim HS, Lee CR, Im JH, Lee KB, Moehl T, Marchioro A, et al. Lead iodide perovskite sensitized all-solid-state submicron thin film mesoscopic solar cell with efficiency exceeding 9%. Scientific Reports. 2012;**2**(1):591

[8] Heo JH, Im SH, Noh JH, Mandal TN, Lim CS, Chang JA, et al. Efficient inorganic–organic hybrid heterojunction solar cells containing perovskite compound and polymeric hole conductors. Nature Photonics. 2013;**7**(6):486-491

[9] NREL best research-cell efficiency chart [Internet]. 2024. Available from: https://www.nrel.gov/pv/assets/pdfs/best-research-cell-efficiencies.pdf [Accessed: July 22, 2024]

[10] Wu WQ, Chen D, Caruso RA, Cheng YB. Recent progress in hybrid perovskite solar cells based on n-type materials. Journal of Materials Chemistry A. 2017;**5**:10092-10109

[11] Foo S, Thambidurai M, Senthil Kumar P, Yuvakkumar R, Huang Y, Dang C. Recent review on electron transport layers in perovskite solar cells. International Journal of Energy Research. 2022;**46**(15):21441-21451

[12] Conings B, Baeten L, Jacobs T, Dera R, D'Haen J, Manca J, et al. An easy-to-fabricate low-temperature TiO_2 electron collection layer for high efficiency planar heterojunction perovskite solar cells. APL Materials. 2014;**2**(8):081505-0815013

[13] Kogo A, Numata Y, Ikegami M, Miyasaka T. Nb_2O_5 blocking layer for high open-circuit voltage perovskite solar cells. Chemistry Letters. 2015;**44**(6):829-830

[14] Liu X, Tsai KW, Zhu Z, Sun Y, Chueh CC, Jen AKY. A low-temperature, solution processable tin oxide electron-transporting layer prepared by the dual-fuel combustion method for efficient perovskite solar cells. Advanced Materials Interfaces. 2016;**3**(13):1600122

[15] Kumar MH, Yantara N, Dharani S, Grätzel M, Mhaisalkar S, Boix PP, et al. Flexible, low-temperature, solution processed ZnO-based perovskite solid state solar cells. Chemical Communications. 2013;**49**:11089-11091

[16] Liu D, Kelly TL. Perovskite solar cells with a planar heterojunction structure prepared using room-temperature solution processing techniques. Nature Photonics. 2013;**8**:133-138

[17] Wang K, Shi Y, Li B, Zhao L, Wang W, Wang X, et al. Amorphous inorganic electron-selective layers for efficient perovskite solar cells: Feasible strategy towards room-temperature fabrication. Advanced Materials. 2016;**28**:1891-1897

[18] Ke W, Fang G, Liu Q, Xiong L, Qin P, Tao H, et al. Low-temperature solution-processed tin oxide as an alternative electron transporting layer for efficient perovskite solar cells. Journal of the American Chemical Society. 2015;**137**:6730-6733

[19] Wang K, Shi Y, Dong Q, Li Y, Wang S, Yu X, et al. Low-temperature and solution-processed amorphous WO_X as electron-selective layer for perovskite solar cells. Journal of Physical Chemistry Letters. 2015;**6**:755-759

[20] Yu J, Li Q, Fan J, Cheng B. Fabrication and photovoltaic performance of hierarchically titanate tubular structures self-assembled by nanotubes and nanosheets. Chemical Communications. 2011;**47**:9161-9163

[21] Zhu X, Cheng B, Yu J, Ho W. Halogen poisoning effect of Pt-TiO_2 for formaldehyde catalytic oxidation performance at room temperature. Applied Surface Science. 2016;**364**:808-814

[22] Lv K, Yu J, Fan J, Jaroniec M. Rugby-like anatase titania hollow nanoparticles with enhanced photocatalytic activity. CrystEngComm. 2011;**13**:7044-7048

[23] Fan J, Zhao L, Yu J, Liu G. The effect of calcination temperature on the microstructure and photocatalytic activity of TiO_2-based composite nanotubes prepared by an in-situ template dissolution method. Nanoscale. 2012;**4**:6597-6603

[24] Scarpelli F, Mastropietro TF, Poerio T, Godbert N. Mesoporous TiO_2 thin films: State of the art. Titanium Dioxide-Material for a Sustainable Environment. 2018;**508**(1):135-142

[25] Verma R, Gangwar J, Srivastava AK. Multiphase TiO_2 nanostructures: A review of efficient synthesis, growth mechanism, probing capabilities, and applications in bio-safety and health. RSC Advances. 2017;**7**(70):44199-44224

[26] Brinker CJ, Scherer GW. Sol-Gel Science: The Physics and Chemistry of Sol-Gel Processing. London, United Kingdom: Academic Press; 2013

[27] Marjunus R, Febriyanti N, Stevani A, Handayani YN, Firdaus I, Manurung P. Synthesis and characterization of TiO2 thin film based on iron sand of Lampung Province-Indonesia. Journal of Physics: Conference Series. 2021;**1951**(1):12002

[28] Liang C, Wu Z, Li P, Fan J, Zhang Y, Shao G. Chemical bath deposited rutile TiO_2 compact layer toward efficient planar heterojunction perovskite solar cells. Applied Surface Science. 2017;**391**:337-344

[29] Ma S, Ahn J, Oh Y, Kwon HC, Lee E, Kim K, et al. Facile Sol gel-derived craterlike dual-functioning TiO_2 electron transport layer for high-efficiency perovskite solar cells.

ACS Applied Materials & Interfaces. 2018;**10**:14649-14658

[30] Yanagisawa K, Ovenstone J. Crystallization of anatase from amorphous titania using the hydrothermal technique: Effects of starting material and temperature. The Journal of Physical Chemistry B. 1999;**103**(37):7781-7787

[31] Shan GB, Demopoulos GP. The synthesis of aqueous-dispersible anatase TiO_2 nanoplatelets. Nanotechnology. 2009;**21**(2):025604

[32] Liu B, Sun G, Sun Q, Lv Y, Huang M, Qi B. Low-temperature fabrication of perovskite solar cells using modified TiO_2 electron transport layer. Materials Science in Semiconductor Processing. 2022;**138**:106303

[33] Supraja S, Chundi N, Ramasamy E, Shanmugasundaram S, Veerappan G. Influence of bi-phasic TiO_2 as a low-temperature curable electron transport layer for efficient perovskite solar cells. Solar Energy. 2022;**247**:308-314

[34] Maryam S, Mufti N, Fuad A, Wisodo H. The effect of photoanode TiO_2/ZnO ratio in perovskite solar cell and its photosensitivity and solar cell performance. In: IOP Conference Series: Materials Science and Engineering. Vol. 515. United Kingdom: IOP Publishing; 2019. p. 012007

[35] Perednis D, Gauckler LJ. Thin film deposition using spray pyrolysis. Journal of Electroceramics. 2005;**14**:103-111

[36] Sun J, Pascoe AR, Meyer S, Wu Q, Della Gaspera E, Raga SR, et al. Ultrasonic spray deposition of TiO_2 electron transport layers for reproducible and high efficiency hybrid perovskite solar cells. Solar Energy. 2019;**188**:697-705

[37] Gupta A, Talha M. Recent development in modeling and analysis of functionally graded materials and structures. Progress in Aerospace Sciences. 2015;**79**:1-14

[38] Wu JJ, Yu CC. Aligned TiO_2 nanorods and nanowalls. The Journal of Physical Chemistry B. 2004;**108**(11):3377-3379

[39] Shalan AE, Narra S, Oshikiri T, Ueno K, Shi X, Wu HP, et al. Optimization of a compact layer of TiO_2 via atomic-layer deposition for high-performance perovskite solar cells. Sustainable Energy & Fuels. 2017;**1**(7):1533-1540

[40] Chang HC, Twu MJ, Hsu CY, Hsu RQ, Kuo CG. Improved performance for dye-sensitized solar cells using a compact TiO_2 layer grown by sputtering. International Journal of Photoenergy. 2014;**1**:380120

[41] Seo H, Son MK, Kim JK, Shin I, Prabakar K, Kim HJ. Method for fabricating the compact layer in dye-sensitized solar cells by titanium sputter deposition and acid-treatments. Solar Energy Materials and Solar Cells. 2011;**95**(1):340-343

[42] Ke W, Fang G, Wang J, Qin P, Tao H, Lei H, et al. Perovskite solar cell with an efficient TiO_2 compact film. ACS Applied Materials & Interfaces. 2014;**6**(18):15959-15965

[43] Jarwal DK, Kumar A, Mishra AK, Ratan S, Kumar C, Upadhyay D, et al. Efficiency improvement of TiO_2 Nanorods electron transport layer based perovskite solar cells by Solvothermal etching. IEEE Journal of Photovoltaics. 2019;**9**:1699-1707

[44] Lu H, Zhong J, Ji C, Zhao J, Li D, Zhao R, et al. Fabricating an optimal rutile TiO_2 electron transport layer by delicately tuning $TiCl_4$ precursor solution for high performance perovskite solar cells. Nano Energy. 2020;**68**:104336

[45] Chen Y, Zhang M, Li F, Yang Z. Recent Progress in perovskite solar cells: Status and future. Coatings. 2023;**13**:644

[46] Sanehira Y, Shibayama N, Numata Y, Ikegami M, Miyasaka T. Low-temperature synthesized Nb-doped TiO_2 electron transport layer enabling high-efficiency perovskite solar cells by band alignment tuning. ACS Applied Materials & Interfaces. 2020;**12**:15175-15182

[47] Culu A, Kaya IC, Sonmezoglu S. Spray-pyrolyzed tantalium-doped TiO_2 compact electron transport layer for UV photostable planar perovskite solar cells exceeding 20% efficiency. ACS Applied Energy Materials. 2022;**5**:3454-3462

[48] Amalathas AP, Landova L, Conrad B, Holovsky J. Concentration-dependent impact of alkali Li metal doped mesoporous TiO_2 electron transport layer on the performance of $CH_3NH_3PbI_3$ perovskite solar cells. Journal of Physical Chemistry C. 2019;**123**:19376-19384

[49] Jin J, Li H, Bi W, Chen C, Zhang B, Xu L, et al. Efficient and stable perovskite solar cells through e-beam preparation of cerium doped TiO_2 electron transport layer, ultraviolet conversion layer CsPbBr3 and the encapsulation layer Al_2O_3. Solar Energy. 2020;**198**:187-193

[50] Arshad Z, Khoja AH, Shakir S, Afzal A, Mujtaba MA, Soudagar MEM, et al. Magnesium doped TiO_2 as an efficient electron transport layer in perovskite solar cells. Case Studies in Thermal Engineering. 2021;**26**:101101

[51] Lv Y, Tong H, Cai W, Zhang Z, Chen H, Zhou X. Boosting the efficiency of commercial available carbon-based perovskite solar cells using zinc-doped TiO_2 nanorod arrays as electron transport layer. Journal of Alloys and Compounds. 2021;**851**:156785

[52] Ren W, Liu Y, Wu Y, Sun Q, Cui Y, Hao Y. Interface modification of an electron transport layer using europium acetate for enhancing the performance of P3HT-based inorganic perovskite solar cells. Physical Chemistry Chemical Physics. 2021;**23**:23818-23826

[53] Sandhu S, Saharan C, Buruga SK, Kumar SA, Rana PS, Nagajyothi PC, et al. Micro structurally engineered hysteresis-free high efficiency perovskite solar cell using Zr-doped TiO_2 electron transport layer. Ceramics International. 2021;**47**:14665-14672

[54] EbrahimiM,KermanpurA,AtapourM, Adhami S, Heidari RH, Khorshidi E, et al. Performance enhancement of mesoscopic perovskite solar cells with GQDs-doped TiO_2 electron transport layer. Solar Energy Materials and Solar Cells. 2020;**208**:110407

[55] Zhang X, Zhang W, Wu T, Wu J, Lan Z. High efficiency and negligible hysteresis planar perovskite solar cells based on NiO nanocrystals modified TiO_2 electron transport layers. Solar Energy. 2019;**181**:293-300

[56] Yue M, Su J, Zhao P, Lin ZH, Zhang JC, Chang JJ, et al. Optimizing the performance of $CsPbI_3$-based perovskite solar cells via doping a ZnO electron transport layer coupled with Interface engineering. Nano-Micro Letters. 2019;**11**:1-14

[57] Zhou J, Lyu M, Zhu J, Li G, Li Y, Jin S, et al. SnO_2 quantum dot-modified mesoporous TiO_2 electron transport layer for efficient and stable perovskite solar cells. ACS Applied Energy Materials. 2022;**5**:3052-3063

[58] Mohseni HR, Dehghanipour M, Dehghan N, Tamaddon F, Ahmadi M, Sabet M, et al. Enhancement of the photovoltaic performance and the

stability of perovskite solar cells via the modification of electron transport layers with reduced graphene oxide/polyaniline composite. Solar Energy. 2021;**213**:59-66

[59] You Y, Tian W, Min L, Cao F, Deng K, Li L. TiO_2/WO_3 bilayer as electron transport layer for efficient planar perovskite solar cell with efficiency exceeding 20%. Advanced Materials and Interfaces. 2020;**7**:1901406

[60] Zhao Y, Zhang H, Ren XG, Zhu HL, Huang ZF, Ye F, et al. Thick TiO_2-based top electron transport layer on perovskite for highly efficient and stable solar cells. ACS Energy Letters. 2018;**3**:2891-2898

[61] Malekshahi Byranvand M, Nemati Kharat A, Fatholahi L, Malekshahi Beiranvand Z. A review on synthesis of nano-TiO_2 via different methods. Journal of Nanostructures. 2013;**3**(1):1-9

[62] Tacchini I, Ansón-Casaos A, Yu Y, Martínez MT, Lira-Cantu M. Hydrothermal synthesis of 1D TiO_2 nanostructures for dye sensitized solar cells. Materials Science and Engineering: B. 2012;**177**(1):19-26

[63] Park YJ, Jeon YI, Yang IS, Choo H, Suh WS, Ju SY, et al. Selective control of novel TiO_2 nanorods: Excellent building blocks for the electron transport layer of mesoscopic perovskite solar cells. ACS Applied Materials & Interfaces. 2023;**15**(7):9447-9456

[64] Elseman AM, Zaki AH, Shalan AE, Rashad MM, Song QL. TiO_2 nanotubes: An advanced electron transport material for enhancing the efficiency and stability of perovskite solar cells. Industrial & Engineering Chemistry Research. 2020;**59**(41):18549-18557

[65] Jiang Q, Sheng X, Li Y, Feng X, Xu T. Rutile TiO_2 nanowire-based perovskite solar cells. Chemical Communications. 2014;**50**(94):14720-14723

[66] Asadzade A, Miandoab SA. Design and simulation of 3D perovskite solar cells based on titanium dioxide nanowires to achieve high-efficiency. Solar Energy. 2021;**228**:550-561

[67] Chen Y, Li L, Quanlong X, Chen W, Dong Y, Fan J, et al. Recent advances in opal/inverted opal, photonic crystal photocatalysts. RRL. 2021;**5**:2000541

[68] Joannopoulos JD, Johnson SG, Winn JN, Meade RD. Molding the Flow of Light. Princeton, NJ [ua]: Princeton University Press; 2008

[69] von Freymann G, Kitaev V, Lotsch BV, Ozin GA. Bottom-up assembly of photonic crystals. Chemical Society Reviews. 2013;**42**(7):2528-2554

[70] Olasoji AJ, Heo JH, Im SH. Facile fabrication of crack-free TiO_2 inverse opal thin-film and its application as electron transporting scaffold for efficient Sb_2S_3-sensitized solar cells. Journal of Colloid and Interface Science. 2025;**678**(B):842-853

[71] Wijnhoven JE, Vos WL. Preparation of photonic crystals made of air spheres in titania. Science. 1998;**281**(5378):802-804

[72] Ha SJ, Heo JH, Im SH, Moon JH. Mesoscopic $CH_3NH_3PbI_3$ perovskite solar cells using TiO_2 inverse opal electron-conducting scaffolds. Journal of Materials Chemistry A. 2017;**5**(5):1972-1977

[73] Wang K, Liu C, Du P, Zheng J, Gong X. Bulk heterojunction perovskite hybrid solar cells with large fill factor. Energy & Environmental Science. 2015;**8**(4):1245-1255

[74] Qin P, Domanski AL, Chandiran AK, Berger R, Butt HJ, Dar MI, et al. Yttrium-substituted nanocrystalline TiO_2 photoanodes for perovskite-based heterojunction solar cells. Nanoscale. 2014;**6**(3):1508-1514

[75] Zhang X, Bao Z, Tao X, Sun H, Chen W, Zhou X. Sn-doped TiO_2 nanorod arrays and application in perovskite solar cells. RSC Advances. 2014;**4**(109):64001-64005

[76] Manseki K, Ikeya T, Tamura A, Ban T, Sugiura T, Yoshida T. Mg-doped TiO_2 nanorods improving open-circuit voltages of ammonium lead halide perovskite solar cells. RSC Advances. 2014;**4**(19):9652-9655

[77] Giordano F, Abate A, Correa Baena JP, Saliba M, Matsui T, Im SH, et al. Enhanced electronic properties in mesoporous TiO_2 via lithium doping for high-efficiency perovskite solar cells. Nature Communications. 2016;**7**(1):10379

[78] Yoo Y, Seo G, Park HJ, Kim J, Jang J, Cho W, et al. Low-temperature rapid UV sintering of sputtered TiO_2 for flexible perovskite solar modules. Journal of Materials Chemistry A. 2024;**12**(3):1562-1572

[79] Yang T, Gao L, Lu J, Ma C, Du Y, Wang P, et al. One-stone-for-two-birds strategy to attain beyond 25% perovskite solar cells. Nature Communications. 2023;**14**(1):839

[80] Jeong J, Kim M, Seo J, Lu H, Ahlawat P, Mishra A, et al. Pseudo-halide anion engineering for α-FAPbI3 perovskite solar cells. Nature. 2021;**592**(7854):381-385

Chapter 8

Perspective Chapter: Titanium Dioxide as a Photocatalysts for Pharmaceutical Waste Degradation Present in Water

Carolina Solis Maldonado, Raúl Alejandro Luna Sánchez, Alfredo Cristobal Salas, Tatiana L. Izaguirre Gallegos, Nayeli Ortiz Silos and José Luis Xochihua Juan

Abstract

This chapter presents a general overview of titanium dioxide (TiO_2) as a key component in photocatalytic degradation processes with emphasis on water quality improvement focusing on the degradation of organic, toxic, persistent, and water-diluted pharmaceutical waste molecules. They are often linked to emerging contaminants such as drug residues and byproducts from significant anthropogenic activities. Throughout the text, there is an emphasis on physical, chemical, and optical properties of titanium dioxide that make it essential in photocatalytic applications. It also introduces fundamental concepts and principles of photocatalysis, facilitating the analysis of optimal experimental parameters and conditions for efficient degradation processes. Additionally, the interaction between titanium dioxide and some pharmaceuticals is examined to understand potential side effects that may impact its efficacy. Finally, the chapter discusses current challenges in the study of titanium dioxide as part of the photocatalytic processes for pharmaceutical waste. Overcoming these challenges would enhance its performance while ensuring its widespread implementation in water treatment systems.

Keywords: photocatalyst, photodegradation, oxidation, TiO_2, pharmaceutical, wastewater, emerging contaminants

1. Introduction

Water quality issues vary significantly across countries depending on regional economic and social conditions. Anthropogenic activities such as agriculture, livestock farming, urbanization, and industrial processes have contributed to the presence of toxic molecules in the environment over time. These emissions are often referred to as emerging contaminants (ECs) when they are deposited in water bodies. The increasing frequency and concentration of these contaminants have become a global concern due to their potential for bioaccumulation [1]. These contaminants pose significant risks to both living organisms and the environment even at low concentrations (μg/L or ng/L). Conventional water treatment methods often fail to address or completely eliminate ECs [2, 3]. Consequently, wastewater treatment plants (WWTP) are usually identified as the major sources of pollution. This is because they release these contaminants into aquatic systems such as oceans, rivers, and streams [4]. Researchers have focused on identifying the primary ECs present in municipal discharges [5, 6]. These studies aim to evaluate the impact of these contaminants on water, sediments, and sludge, as well as to assess the quality of major water bodies posttreatment by WWTP. Among the main ECs identified in municipal waters are industrial and household chemicals, including artificial sweeteners [7, 8], personal care product ingredients [9, 10], pharmaceuticals [11, 12], hormones [13, 14], UV filters [15, 16] pesticides [17, 18], plasticizers, and flame retardants [19]. More sensitive analytical technologies [20, 21] have facilitated the detection of these contaminants, raising concerns about their potential environmental and health implications [22, 23].

In recent years, the development of various technologies aiming to minimize or degrade emerging contaminants (EC) has increased. These new technologies have become a complementary approach to conventional physical, chemical, or biological treatment carried out in WWTP. These posttreatments are based on adsorption processes [24, 25] and on the degradation of molecules through advanced oxidation processes [26], using nanoparticles or nanomaterials as adsorbents [27] or photocatalysts [28, 29]. Titanium oxide (TiO_2) is being widely used due to its efficiency in the degradation of ECs when it is exposed to ultraviolet (UV) light [30, 31]. Currently, research is oriented to improve photocatalysts to promote further development of its application in water treatment [32] and its implementation in WWTPs. In recent years, various interest groups have promoted research on wastewater treatments [33, 34], seeking an innovative, efficient, economic, and low-energy alternative for use at a domestic scale. These initiatives intend to ensure implementation and maintenance of technological infrastructure that meets the water consumption requirements. Therefore, this chapter presents the environmental impact caused by the presence of pharmaceutical waste and its byproducts in water bodies. Properties that characterize a photocatalyst are also presented in the following sections. Also, topics such as water pollution, titanium oxide (TiO_2) as a photocatalyst, and catalytic photodegradation of pharmaceuticals are also analyzed. Finally, some advantages of TiO_2 as a catalyst are addressed. The chapter closes analyzing the challenges and perspectives of this element as a "new-proposed stage" in water treatment plants.

2. Water pollution

Water pollution by ECs is becoming an environmental problem of global relevance. These contaminants have multiple and varied origins, including personal care products, agrochemicals, hormones, textile dyes, fire retardants and pharmaceuticals, among others [35]. In addition, these contaminants can interact with organic matter; they are soluble in water and persistent in the environment and they are bioaccumulative. Even at low concentrations (from ngL −1 a μgL −1), they can significantly affect the ecosystem [36]. In particular, pharmaceuticals represent an important problem that requires immediate action. Their chemical stability, their extensive global consumption, their incessant arrival into the environment, and their difficult degradation through biodegradable processes or conventional technologies place them among the emerging contaminants of interest for some researchers [37]. Additionally, emerging contaminants derived from pharmaceutical products represent a growing concern due to their negative impacts on the environment and human health [38]. For example, hormonal compounds can alter the endocrine system of fish resulting in phenomena such as feminization of males and changes in reproduction and growth patterns [39, 40]. Another example occurs when the presence of antibiotics in the aquatic environment can promote the appearance and proliferation of resistant bacterial strains, complicating the treatment of infections in humans and animals [41]. The bioaccumulation of pharmaceutical products affects the food chain of fish, which can cause risks to human health [42]. A worrying aspect of this problem is the presence of pharmaceutical products and their transformation products in various surfaces, underground, residual, and drinking water sources. These products enter the aquatic environment through multiple sources, including landfills, incineration, pharmaceutical waste from livestock, domestic waste, and hospital and industrial waste (**Figure 1**). These primary sources generate waste products that affect the soil, water, and atmosphere. Contaminants can filter into groundwater after passing through various soil layers and treatment systems [43].

Most of the emerging contaminants are not easily biodegradable and have complex chemical structures that allow them to pass through the filtration processes of WWTPs without being effectively removed [44, 45]. That is why advanced oxidative processes (AOPs) are proposed for the treatment of these contaminants. These processes include techniques such as electrochemical processes, ozonation, photolysis, the Fenton process, photo-Fenton, and photocatalysis, among others. These methods are proven to be highly effective for the degradation of molecules with characteristics like those of emerging contaminants. This effectiveness lies in their ability to generate highly reactive free radicals, such as the hydroxyl radical (•OH). These methods can break down the complex chemical structures of these compounds and transform them into less toxic and simpler end products such as carbon dioxide and water [46, 47].

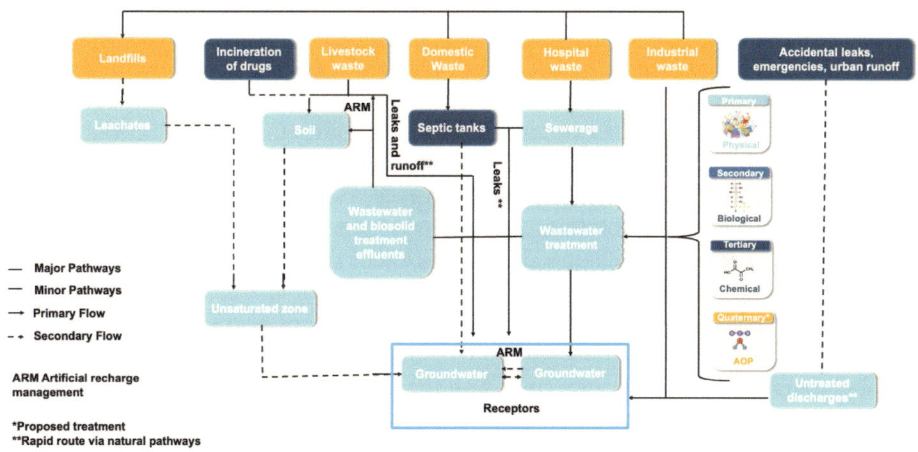

Figure 1.
Routes of entry of medications into aquatic environments.

3. Titanium oxide (TiO_2) as photocatalyst

Catalysis is a field of chemistry that studies changes in the rate of chemical reactions. These changes are attributed to the incorporation of a material, called a catalyst, which is added in a chemical reaction with the intention of transforming one molecule into another one. Catalysts must have physical and chemical properties that allow the acceleration of the transformation process. Physical properties include surface area, porosity, pore size, and volume. These properties allow the incorporation and availability of the active sites, as well as the circulation of the reactants. The chemical properties are the chemical composition, the active sites, and acidity. The active sites favor lower activation energy (Eq. 1), to reach the transition state necessary for the reaction to occur. These characteristics make the catalyst specific, special, or unique for certain reactions and stable under different conditions. This is why chemical and physical properties define the catalytic performance of the catalyst in a reaction [48]. On the other hand, catalysts must also have catalytic properties such as activity, selectivity, and stability. The activity of a catalyst refers to its ability to accelerate a chemical reaction. This property is measured as a rate where the reaction progresses in the presence of the catalyst [49]. Also, the Turnover Number (TON, Eq. 2) and Turnover Frequency (TOF, Eq. 3) are used to evaluate its effectiveness. These frequencies allow the comparison of catalytic performance with other catalysts under different conditions. These measurements consider the number of active molecules converted per catalytic active site and per second [50]. If the catalyst does not present activity in a particular reaction, even if it apparently has certain chemical or physical properties, it cannot be considered a good catalyst.

$$-r_A = \frac{Amount\ of\ substance\ converted\,(mol)}{mass\ of\ catalyst\,(g) \times time\,(s)} \qquad (1)$$

$$ON = \frac{number\ of\ moles\ of\ product\ obtained}{moles\ of\ catalyst} \qquad (2)$$

$$TOF = \frac{number\ of\ moles\ of\ product\ obtained}{time \times moles\ of\ catalyst} \qquad (3)$$

The selectivity property of a catalyst refers to its ability to favor the production of a specific product. This production can be carried out at a higher rate or in greater quantity over other undesired products that can also be formed during a reaction. This property is crucial for obtaining specific products with high purity and depends on the structure, phase, and chemical composition of the catalyst, specifically on the active sites that interact with the reagents. The property of chemical stability during the reaction of a catalyst is also crucial. Catalysts should not degrade, transform, undergo structural alterations, or accumulate impurities under the reaction conditions (temperature, pH, agitation, time, and toxic substances, among others). The deactivation of the catalyst and the loss of catalytic activity are favored by these factors. On the contrary, a good catalyst must be stable and capable of being regenerated after each reaction cycle; that is, it must be able to be recovered and reused with the same efficiency.

Catalysis can be carried out in two ways: homogeneous or heterogeneous. Homogeneous catalysis is distinguished by the fact that the catalyst and the reactants are in the same phase, either in the liquid phase (most commonly used) or in the gas phase. In heterogeneous catalysis, the catalyst is in a different phase than the reactants. For example, it is common for the catalyst to be in the solid phase, while the reactants can be in the liquid or gas phase. In general, these two approaches have their advantages and disadvantages when carrying out a chemical reaction (**Table 1**). In the case of homogeneous catalysis, the main advantage is that there is a uniform distribution when it is in the same phase and also a greater contact with the reagents. This distribution favors the efficiency of the reaction and obtains a better catalytic activity as well as greater selectivity. Meanwhile, heterogeneous catalysis has the advantage of its easy separation and recovery of the catalyst. This favors the purity of products from contaminating substances or molecules [51].

During the chemical reaction, the catalyst is in direct contact with the reactants and activated to facilitate or accelerate the reaction. Catalysts can also be classified by the element that activates them. For example, catalyst activation can be carried out by exposing the mixture to a specific temperature or light source. Activation is mostly promoted by a thermal process. The temperature required for this to occur can vary and depends on the chemical molecule or active site to be activated [52]. On the other hand, there are catalysts that are activated when exposed to light (photons), mainly ultraviolet (UV), visible (VIS), or solar light. Photocatalysts are activated when exposed to photons of light. These photons excite electrons on the surface of the material, creating electron-hole pairs (negative electrons and positive holes). During excitation, reactive species (free radicals) are produced to participate in the reactions, favoring the breakdown of organic molecules. The effective activation of a photocatalyst depends on the type of light (wavelength, intensity, and duration of exposure). For a photocatalyst to have good catalytic activity, it must have certain optical properties that favor the interaction of light with the active sites. The band gap is one of the important properties of photocatalysts since it defines the wavelengths of light that can be reflected, scattered, or absorbed in the material, playing an important role in the excitation of the active sites [53].

Photocatalysis is an advanced oxidation process (**Figure 2**). During the reaction, the photocatalyst promotes the formation of reactive species (hydroxyl radicals) that interact with reactants (organic molecules). The photocatalyst first absorbs photons to form the electron-hole pair and then promotes the migration of the electron-hole pair to the catalyst surface. Then, it interacts with the water molecules (H_2O) and oxygen (O_2), oxidizing water to form the hydroxyl radical

Features	Homogeneous	Heterogeneous
Activity	High	Variable
Selectivity	High	Moderate
Stability	Low	High
Recovery	Difficult	Easy
Reusable	Low	High

Table 1.
Advantages and disadvantages of homogeneous and heterogeneous catalysis.

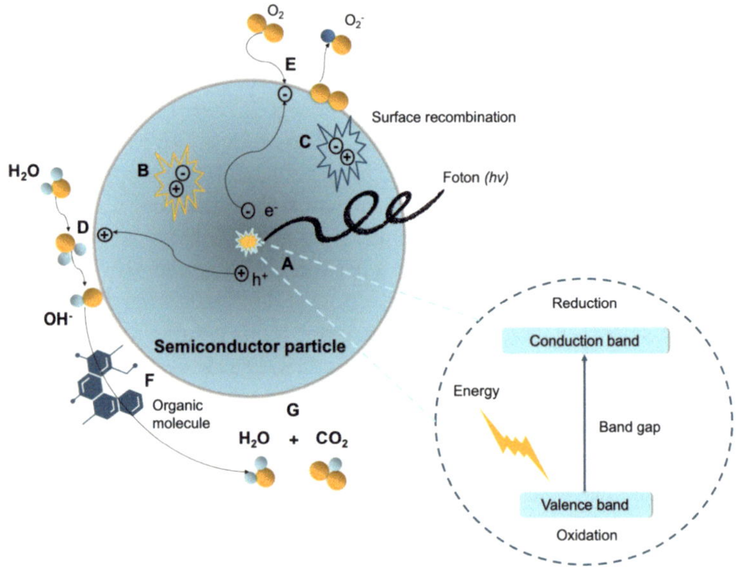

Figure 2.
Photocatalyst mechanism diagram.

(•OH) or reducing oxygen to form the superoxide radical (O^{-2}), respectively. Hydroxyl radicals are very reactive in the presence of organic molecules (e.g., contaminant molecules), until mineralization can be achieved, forming molecules such as carbon dioxide (CO_2) and water (H_2O) [54, 55]. Titanium oxide (TiO_2) is the most widely used material as a photocatalyst. This material has ideal properties to be activated in the presence of photons. The main properties of this material are its optical characteristics, which classify it as a good semiconductor in the presence of light. TiO_2 has a band gap of approximately 3.0 eV for the anatase phase and 3.2 eV for the rutile phase, active within this range of ultraviolet (UV) light, and generates electron-hole pairs necessary for photocatalysis. Additionally,

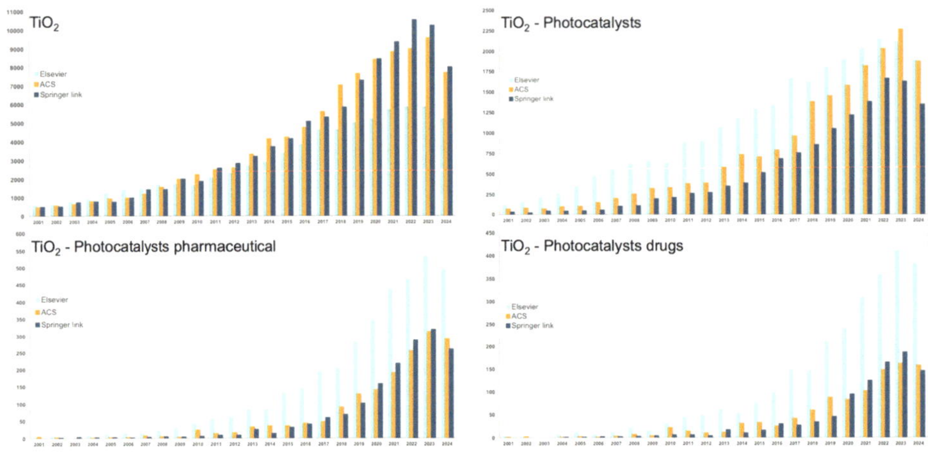

Figure 3.
Publication of research works related to titanium oxide [58–60].

TiO_2 is stable and resistant in corrosive environments, nontoxic, and naturally abundant, which makes it economical, with good performance throughout the process [56, 57].

In the last 20 years, many research works are related to the use of titanium oxide (**Figure 3**). These studies range from the design of the photocatalytic material or its improvement, to its application in the degradation or conversion of molecules, and only 22% are focused on its application as a photocatalyst. As explained, photocatalysts, due to their optical properties, are ideal materials for the degradation of toxic organic molecules classified as emerging contaminants. Therefore, the number of studies related to the application of TiO_2 as a photocatalyst in the degradation of emerging molecules has also increased, and only in the case of pharmaceuticals, it is around 5%. This result represents a large area of opportunity that should be developed.

4. Catalytic photodegradation of pharmaceuticals

Pharmaceuticals are chemical compounds used for the treatment and prevention of diseases in humans and animals. These compounds are designed to be biologically active and persistent, ensuring their therapeutic efficacy. Their classification is wide and includes antibiotics, analgesics, antifungals, anticonvulsants, beta blockers, hormones, and lipid regulators, among others [61, 62]. Monitoring studies detected the presence of pharmaceutical compounds in the environment, resulting from the inappropriate excretion and disposal of medicines by consumers [63]. The literature reports their presence in various bodies of water, such as surface water, underground water, drinking water, and affluents from wastewater treatment plants. This phenomenon varies according to the region, time of year, climatic conditions, and consumption patterns of each country. Among the main compounds detected are acetaminophen, ibuprofen, diclofenac, naproxen, codeine, gemfibrozil, sulfamethoxazole, erythromycin, triclosan, trimethoprim, metoprolol, and carbamazepine. These compounds appear in low concentrations, ranging from ng/L to g/L, and their presence has been increasing in recent decades [64]. The negative effects of drugs and their metabolites on the ecosystem are attributed to their physicochemical properties, such as high polarity, volatility, lipophilicity, and adsorption capacity [65]. Their continuous release into the aquatic environment through various pathways, such as treated or untreated wastewater from municipal, hospital, and industrial plants, landfill leachates, and surface runoff from urban or agricultural areas, classifies them as emerging and pseudopersistent microcontaminants [66, 67]. The persistence of these compounds in the environment is due to their resistance to degradation by natural processes, and conventional treatments are not effective in reducing their levels or eliminating them. The potential risk of pharmaceutical compounds at environmental concentrations on aquatic organisms and human health is not yet fully understood. However, studies have revealed that even at trace concentrations, many pharmaceuticals cause negative effects, such as reproductive alterations due to hormonal activity, diabetes, obesity, endometriosis, and antibiotic resistance, among others [68].

Removing or reducing molecules from pharmaceuticals presents a significant challenge due to the complexity of environmental mixtures and matrices [69]. Experimental parameters such as the amount of catalyst, the concentration of the pharmaceutical compound, the presence of electron acceptors, and the pH are crucial to maximize the removal and mineralization efficiency of the pharmaceuticals, and their intermediates generated during the process [70]. Research studies showed a relationship between the percentage of degradation and the concentration of pharmaceutical compounds, attributed to the saturation of the active sites on the TiO_2 surface by high concentrations of organic molecules and their byproducts, limiting the sites available for the generation of hydroxyl radicals [71]. The mass of TiO_2 also influences the percentage of degradation of pharmaceutical compounds. Although a higher catalyst loading provides more surface area for adsorption of molecules, a screening effect can occur in the center of the reactor, where internal particles do not contribute to electron-hole pair production due to the excess of TiO_2 [72]. This phenomenon has been observed in the case of paracetamol, where the optimal photocatalyst load was determined to avoid this effect under specific conditions [73]. pH is another factor that can affect molecule degradation rates with TiO_2. The photocatalyst surface can be positively charged in solutions with pH below 6 (zero charge point of TiO_2) or negatively charged above this value, which influences the adsorption of pharmaceuticals on the TiO_2 surface. It is essential to monitor the pH behavior in the degradation of specific molecules, as it affects its photocatalysis [74]. In addition, the composition of TiO_2 plays an important role in its photocatalytic activity. The anatase phase has been shown to degrade clofibric acid more quickly than rutile catalysts and the commercial Aeroxide P_{25}, which contains mixed fractions of both phases [75]. Many research studies indicate that commercial TiO_2 P_{25} (70:30, anatase:rutile) degradation is more efficient due to reduced recombination of electron-hole pairs on the surface or bulk of the photocatalyst in the presence of UV light [76].

TiO_2 doping is an effective strategy to reduce the band gap, allowing its activation with lower energy and facilitating the photocatalysis of pharmaceuticals under visible light [77]. The radiation source and intensity used to activate the photocatalyst are also crucial factors [78, 79]. The photodegradation mechanisms of drugs with TiO_2 have been widely described. In some cases, it has been reported that the products generated by oxidation may have a higher toxicity than the original compound, as in the case of diclofenac and ibuprofen; in other cases, complete mineralization of the drugs has been demonstrated [80, 81]. Regarding the efficiency and conditions of drug degradation using exclusively TiO_2 as a photocatalyst, pharmaceuticals are classified into groups: analgesics and anti-inflammatories, antibiotics, antiepileptics, antidepressants, lipid-lowering agents, and beta-blockers, among others [82]. Nonsteroidal anti-inflammatory drugs (NSAIDs) are a group of molecules used to reduce fever, pain, and inflammation. Their high consumption has turned them into emerging microcontaminants in bodies of water. The degradation of ibuprofen, naproxen, diclofenac, acetaminophen (paracetamol), acetylsalicylic acid (aspirin), ketoprofen, and metamizole, among others, has been widely studied in the presence of TiO_2, varying the process conditions, obtaining kinetic parameters of the degradation rates, and considering them as model molecules to test new photocatalysts [83].

Antibiotics are designed to reduce microbes, safeguarding human and animal life. Examples of these antibiotics include azithromycin, sulfamethoxazole, ofloxacin, trimethoprim, ampicillin, doxycycline, tylosin, sulfathiazole, chloramphenicol, vancomycin, amoxicillin, cloxacillin, metronidazole, tetracycline, ciprofloxacin, moxifloxacin, clarithromycin, and erythromycin. Photocatalysis with TiO_2 is an efficient method to degrade these compounds in water. However, the byproducts generated must be tested for ecotoxicity [84]. During the degradation of antiepileptic drugs such as carbamazepine and antidepressants such as benzodiazepine, venlafaxine, sertraline, and promazine, the porous structure of the photocatalyst has been observed to improve the degradation efficiency. The kinetic data match the first-order reaction rate law during the first few minutes of irradiation [85]. Lipid control drugs, which lower plasma cholesterol and triglyceride levels, have also been studied in photocatalysis. The family of fibrates (bezafibrate, fenofibrate, and gemfibrozil) and atorvastatin and simvastatin have shown complete degradation using solar simulators to activate TiO_2, reporting pseudo–first-order kinetics [67]. Beta blockers, used to reduce blood pressure, include compounds such as metoprolol, propranolol, and atenolol. High percentages of mineralization have been reported due to the attack of hydroxyl radicals on the aromatic benzene ring of these molecules in the presence of commercial TiO_2 and UV light [63]. In addition, contraceptives, antidiabetics, stimulants, and opiates have been studied due to their presence in wastewater, developing scientific interest in their elimination through advanced oxidation processes [61]. These studies clearly suggest that organic molecule degradation processes in aqueous solution using TiO_2 are successful. Parameters such as initial drug concentration, catalyst dosage, pH, catalyst nature (doped/undoped), light source, stirring speed, and photoreactor design are crucial to achieve high efficiency and complete oxidation of drugs into less polluting products [86].

5. Advantages, challenges, and prospects

The management and treatment of various environmental contaminants have driven the development of sustainable proposals and strategies. One of the most promising alternatives in environmental technology is AOPs, specifically heterogeneous photocatalysis [87]. This technology offers significant advantages compared to conventional wastewater treatments [88, 89]. The remarkable acceptance of TiO_2 is due to its large surface area, adsorption capacity, photostability [90], chemical and thermal stability, low cost, environmental friendliness and, mainly, its high photocatalytic activity [91]. However, there are some important limitations in the application of TiO_2 as a photocatalyst, which directly affect its performance. These limitations are related to the properties of the photocatalyst and engineering aspects of the design of water treatment equipment [92]. The crystalline structure of TiO_2, specifically the anatase phase, provides a higher surface area [93] and greater photocatalytic efficiency. However, the incidence of light photons with wavelengths less than 400 nm is essential; that is, TiO_2 specifically requires ultraviolet (UV) light energy for its proper activation [94]. This necessity represents a limitation for the use of TiO_2 in photocatalytic processes under solar radiation, since TiO_2 can absorb approximately only 5% of the energy coming from solar radiation, corresponding to the UV light region. Therefore, the rest of the visible light energy (~43%) and infrared light energy (~52%) is not used, which does not provide an additional benefit in chemical transformations (redox reactions) or in the photocatalytic process [95].

Another aspect to consider is the high percentage of recombination of charge carriers (electron-hole pairs) within or on the surface of the photocatalyst. The recombination rate can reach up to 90%, which counteracts the expected efficiency of the photocatalytic process. The efficiency of photocatalysis can be affected by the concentration of contaminants and the adsorption of inorganic ions on the surface of the photocatalyst. These contaminants can compete with the photocatalyst for the absorption of light photons, necessary for its activation and photocatalytic performance, as previously mentioned. Similarly, the adsorption of inorganic ions (phosphates, nitrates, and/or sulfates) on the surface of the photocatalyst can block its active sites and form a thin film that hinders the effective adsorption of the contaminants to be degraded [92]. From a technical approach, establishing a wastewater treatment system using photocatalytic processes (on a pilot or industrial scale) may require a considerable investment due to its maintenance and consumption of chemical reagents. This aspect arises from the need to perform various experimental tests to determine the optimal operating conditions (catalyst, pH, etc.) and obtain the best results in the process. However, the greatest challenge lies in the demand for UV light energy needed for its implementation. To address this energy need, efforts have been directed toward harnessing renewable energy sources, such as solar radiation [96]. In this context, the equipment used must meet the following conditions: favor the capture of solar radiation, inhibit recombination reactions, regulate temperature conditions, and ensure that the investment is the most economical option [97]. It is important to mention that the technical aspects are crucial, but they do not cover all the drawbacks. An additional complication is the complete recovery of the photocatalyst from the treated aqueous medium through a filtration process, since not doing so and releasing it directly could generate a new problem due to the toxicity of TiO_2 at the environmental level [98, 99]. The volumes of treated water obtained through this alternative technology are usually small compared to conventional systems. In response to these areas of opportunity and with the aim of proposing photocatalytic processes as a viable technology for wastewater treatment, various studies have been developed to improve the performance of the photocatalyst [96].

Experimental results indicate the inhibiting the recombination process and enhancing chemical transformations is performed on the surface of the photocatalyst, hence the importance of the synthesis process, where the surface characteristics (physical and chemical) are defined. Regarding the use of solar radiation as a renewable energy source for photocatalytic processes, most strategies focus on improving the physicochemical and optical properties of TiO_2, mainly by increasing the absorption of visible light photons. To achieve this goal, it is necessary to dope and codope TiO_2, as well as the incorporation of dyes to sensitize the surface of the photocatalyst [100]. For the modification of TiO_2 (doping or codoping), various types of ions are used, and most of them come from transition metals (vanadium, nickel, copper, and iron), noble metals (ruthenium, palladium, platinum, gold, and silver), and rare earth metals (cerium, gadolinium, terbium, and europium, among others). These modifications reduce the band gap value of pure TiO_2, allowing its activation with visible light photons (>400 nm) from solar radiation. It has been reported that an additional benefit is the inhibition of electron-hole pair recombination reactions [101]. On the other hand, the incorporation of dyes aims to sensitize the surface of the photocatalyst and increase the absorption of visible light, improving its photoactivation [102]. The most notable equipment is parabolic trough collector reactors (PTC). These reactors have been shown to cover various operational needs, such as solar radiation collection and temperature control. In addition, they have overcome intrinsic physicochemical limitations by minimizing recombination processes, also achieving the degradation of emerging contaminants [103]. These equipment's are used in process of water disinfection, detoxification, and degradation of emerging contaminants derived from pharmaceuticals [104, 105] and pesticides [106]. PTC reactors are promising for water decontamination, although they only allow treating small volumes of water, from 0.4 L to approximately 10 L. Therefore, an industrial-scale wastewater treatment system with PTC reactors would require a large surface area and a high economic investment [107]. It is necessary to assess the technical and economic feasibility of installing these photocatalytic systems.

Another important factor is determining the amount of energy that can be used, since solar energy depends on the geographic location. This technology is more attractive in regions with greater exposure to solar radiation, achieving a greater economic and social impact [108]. Heterogeneous photocatalysis has driven the development of numerous studies at the laboratory level, extending to commercial and industrial applications as a sustainable alternative for the treatment of various contaminants. This type of catalysis also guarantees adequate disposal of the generated products. Therefore, a more practical way of using TiO_2 has been sought, avoiding its recovery by filtration and optimizing work time. One of the best options is the immobilization of commercial TiO_2 powder on different types of substrates. Additionally, borosilicate glass tubes or low-density polyethylene supports are used as necessary elements in this type of catalysis [109]. Usually, the deposition of the photocatalyst (immobilization) on different substrates is recommended to avoid its leaching during the degradation of contaminants in aqueous solution. This helps to counteract the concern about TiO_2 contamination in the environment [110]. The use of TiO_2 can become a crucial element in these advanced oxidation processes directing the sector toward materials science and the application of nanotechnology. Modifications to pure TiO_2 (doping, codoping, sensitization, and immobilization) should pursue compelling benefits, such as the use of solar radiation to position heterogeneous photocatalysis as a sustainable technology [111]. It is essential to respond to the morphological, physicochemical, and optical properties of the photocatalyst in relation to the percentage of degradation of the contaminants. Therefore, it is suggested to carry out the necessary characterizations of these materials and understand the reaction mechanism. It is necessary to know the separation and capture of the charge carriers as well as to look for strategies to prolong their lifetime. There is still much work to be done, from the technical aspects to the intrinsic properties of the photocatalyst. There cannot be immediate responses to all needs, but collaborative and multidisciplinary work by the scientific community is crucial to promote scientific and technological development, contributing to social, economic, and environmental benefits.

6. Conclusion

Titanium oxide is a widely studied material that has proven its effectiveness as a photocatalyst. Over the past twenty years, the number of studies focusing on its performance in the degradation of toxic organic molecules present in water has increased significantly. Greater effort and knowledge from multidisciplinary teams are required for this type of technology to become a reality. It is expected that they can be implemented in wastewater treatment processes and used for their effectiveness in eliminating multiple emerging contaminants.

Conflict of interest

The authors declare no conflict of interest.

Appendices and nomenclature

$-R_A$	activation energy
POA	advanced oxidative processes
CO_2	carbon dioxide
eV	electronvolts
CE	emerging contaminants
•OH	hydroxyl radical
O_2	oxygen
O^{-2}	superoxide radical
TiO_2	titanium oxide
TON	turnover number
TOF	turnover frequency
UV	ultraviolet
VIS	visible

Author details

Carolina Solis Maldonado[1*], Raúl Alejandro Luna Sánchez[1], Alfredo Cristobal Salas[2], Tatiana L. Izaguirre Gallegos[3], Nayeli Ortiz Silos[1] and José Luis Xochihua Juan[1]

1 Faculty of Chemical Sciences, Veracruzana University, Veracruz, Mexico

2 Faculty of Electronics and Communications Engineering, Veracruzana University, Veracruz, Mexico

3 Faculty of Mechanical and Electrical Engineering, Veracruzana University, Veracruz, Mexico

*Address all correspondence to: casolis@uv.mx

References

[1] Gyojin C, Wang W, Cho HS, Kim K, Park K, Oh JE. Legacy and emerging persistent organic pollutants in the freshwater system: Relative distribution, contamination trends, and bioaccumulation. Environment International. 2020;**135**:105377. DOI: 10.1016/j.envint.2019.105377

[2] Poynton HC, Robinson WE. Contaminants of emerging concern, with an emphasis on nanomaterials and pharmaceuticals. In: Török B, Dransfield T, editors. Green Chemistry. Elsevier; 2018. pp. 291-315. ISBN: 9780128092705. DOI: 10.1016/B978-0-12-809270-5.00012-1

[3] Jelic A, Gros M, Ginebreda A, Cespedes-Sánchez R, Ventura F, Petrovic M, et al. Occurrence, partition and removal of pharmaceuticals in sewage water and sludge during wastewater treatment. Water Research. 2011;**45**(3):1165-1176. DOI: 10.1016/j.watres.2010.11.010

[4] Merhaby D, Net S, Halwani J, Ouddane B. Organic pollution in surficial sediments of Tripoli Harbour, Lebanon. Marine Pollution Bulletin. 2015;**93**(1-2):284-293. DOI: 10.1016/j.marpolbul.2015.01.004

[5] Huang YH, Dsikowitzky L, Yang F, Schwarzbauer J. Emerging contaminants in municipal wastewaters and their relevance for the surface water contamination in the tropical coastal city Haikou, China. Estuarine, Coastal and Shelf Science. 2020;**235**:106611. DOI: 10.1016/j.ecss.2020.106611

[6] Dsikowitzky L, Crawford SE, Nordhaus I, Lindner F, Dwiyitno IHE, Ariyani F, et al. Analysis and environmental risk assessment of priority and emerging organic pollutants in sediments from the tropical coastal megacity Jakarta, Indonesia. Regional Studies in Marine Science. 2020;**34**:101021. DOI: 10.1016/j.rsma.2019.101021

[7] Baena-Nogueras RM, Traverso-Soto JM, Biel-Maeso M, Villar-Navarro E, Lara-Martín PA. Sources and trends of artificial sweeteners in coastal waters in the bay of Cadiz (NE Atlantic). Marine Pollution Bulletin. 2018;**135**:607-616. DOI: 10.1016/j.marpolbul.2018.07.069

[8] Ordonez EY, Quintana JB, Rodil R, Cela R. Determination of artificial sweeteners in sewage sludge samples using pressurised liquid extraction and liquid chromatography–tandem mass spectrometry. Journal of Chromatography. A. 2013;**1320**:10-16. DOI: 10.1016/j.chroma.2013.10.049

[9] Pérez-Coyotla I, Galar-Martínez M, García-Medina S, Gómez-Oliván LM, Gasca-Pérez E, Martínez-Galero E, et al. Polluted water from an urban reservoir (Madín dam, México) induces toxicity and oxidative stress in Cyprinus carpio embryos. Environmental Pollution. 2019;**251**:510-521. DOI: 10.1016/j.envpol.2019.04.095

[10] Anekwe JE, Oluseyi T, Drage DS, Harrad S, Abdallah ME. Occurrence, seasonal variation and human exposure to pharmaceuticals and personal care products in surface water, groundwater and drinking water in Lagos State, Nigeria. Emerging Contaminants. 2020;**6**:124-132. DOI: 10.1016/j.emcon.2020.02.004

[11] Huang L, Mao N, Yan Q, Zhang D, Shuai Q. Magnetic covalent organic frameworks for the removal of diclofenac

sodium from water. ACS Applied Nano Materials. 2020;**3**:319-326. DOI: 10.1021/acsanm.9b01969

[12] Chen P, Blaney L, Cagnetta G, Huang J, Wang B, Wang Y, et al. Degradation of ofloxacin by perylene diimide supramolecular nanofiber sunlight-driven photocatalysis. Environmental Science and Technology. 2019;**53**:1564-1575. DOI: 10.1021/acs.est.8b05827

[13] Estrada-Arriaga EB, Cortés-Muñoz JE, González-Herrera A, Calderón-Mólgora CG, Rivera-Huerta MDL, Ramírez-Camperos E, et al. Assessment of full-scale biological nutrient removal systems upgraded with physico-chemical processes for the removal of emerging pollutants present in wastewaters from Mexico. Science of the Total Environment. 2016;**571**:1172-1182. DOI: 10.1016/j.scitotenv.2016.07.118

[14] Lesser LE, Mora A, Moreau C, Mahlknecht J, Hern A. Survey of 218 organic contaminants in groundwater derived from the world's largest untreated wastewater irrigation system: Mezquital. Chemosphere. 2018;**198**:510-521. DOI: 10.1016/j.chemosphere.2018.01.154

[15] Labille J, Slomberg D, Catalano R, Robert S, Apers-Tremelo ML, Boudenne JL, et al. Assessing UV filter inputs into beach waters during recreational activity: A field study of three French Mediterranean beaches from consumer survey to water analysis. Science of the Total Environment. 2020;**706**:136010. DOI: 10.1016/j.scitotenv.2019.136010

[16] Rodil R, Villaverde-de-Sáa E, Cobas J, Quintana JB, Cela R, Carro N. Legacy and emerging pollutants in marine bivalves from the Galician coast (NW Spain). Environment International. 2019;**129**:364-375. DOI: 10.1016/j.envint.2019.05.018

[17] Hernández-Ramírez AG, Martínez-Tavera E, Rodríguez-Espinosa PF, Mendoza-Pérez JA. Detection, provenance and associated environmental risks of water quality pollutants during anomaly events in River Atoyac, Central México: A real-time monitoring approach. Science of the Total Environment. 2019;**669**:1019-1032. DOI: 10.1016/j.scitotenv.2019.03.138

[18] Polanco-Rodríguez AG, Riba-López MI, Del Valls CA, Araujo-León JA, Banik SD. Impact of pesticides in karst groundwater. Groundwater for Sustainable Development. 2018;**7**:20-29. DOI: 10.1016/j.gsd.2018.02.003

[19] Martín J, Zafra-Gómez A, Hidalgo F, Ibañez-Yuste AJ, Alonso E, Vilchez JL. Multi-residue analysis of 36 priority and emerging pollutants in marine echinoderms (Holothuria tubulosa) and marine sediments by solid-liquid extraction followed by dispersive solid phase extraction and liquid chromatography–tandem mass spectrometry analysis. Talanta. 2017;**166**:336-348. DOI: 10.1016/j.talanta.2017.01.062

[20] Álvarez-Ruiz R, Picó Y. Analysis of emerging and related pollutants in aquatic biota. Trends in Environmental Analytical Chemistry. 2020;**25**:e00082, 1-9. DOI: 10.1016/j.teac.2020.e00082

[21] Magi E, Di Carro M, Mirasole C, Benedetti B. Combining passive sampling and tandem mass spectrometry for the determination of pharmaceuticals and other emerging pollutants in drinking water. Microchemical Journal. 2018;**136**:56-60. DOI: 10.1016/j.microc.2016.10.029

[22] Barceló D. Emerging pollutants in water analysis. TrAC Trends in Analytical

Chemistry. 2013;**22**:1-7. DOI: 10.1016/S0165-9936(03)01106-3

[23] Petersen K, Hasle Heiaas H, Tollefsen KE. Combined effects of pharmaceuticals, personal care products, biocides and organic contaminants on the growth of Skeletonema pseudocostatum. Aquatic Toxicology. 2014;**150**:45-54. DOI: 10.1016/j.aquatox.2014.02.013

[24] Flores-Céspedes F, Villafranca-Sánchez M, Fernández-Pérez M. Alginate-based hydrogels modified with olive pomace and lignin for removal of organic pollutants from aqueous solutions. International Journal of Biological Macromolecules. 2020;**153**:883-891. DOI: 10.1016/j.ijbiomac.2020.03.081

[25] Herrera BA, Bruna TC, Sierpe RA, Lang EP, Urzúa M, Flores MI, et al. A surface functionalized with per-(6-amino-6-deoxy)-β-cyclodextrin for potential organic pollutant removal from water. Carbohydrate Polymers. 2020;**233**:115865. DOI: 10.1016/j.carbpol.2020.115865

[26] Loeb SK, Alvarez PJJ, Brame JA, Cates EL, Choi W, Crittenden J, et al. The technology horizon for photocatalytic water treatment: Sunrise or sunset? Environmental Science and Technology. 2019;**53**:2937-2947. DOI: 10.1021/acs.est.8b05041

[27] Lu F, Astruc D. Nanocatalysts and other nanomaterials for water remediation from organic pollutants. Coordination Chemistry Reviews. 2020;**408**:213180. DOI: 10.1016/j.ccr.2020.213180

[28] Dutta K, Rana D. Polythiophenes: An emerging class of promising water purifying materials. European Polymer Journal. 2019;**116**:370-385. DOI: 10.1016/j.eurpolymj.2019.04.033

[29] Subudhi S, Mansingh S, Swain G, Behera A, Rath D, Parida K. HPW-anchored UiO-66 metal–organic framework: A promising photocatalyst effective toward tetracycline hydrochloride degradation and H_2 evolution via Z-scheme charge dynamics. Inorganic Chemistry. 2019;**58**:4921-4934. DOI: 10.1021/acs.inorgchem.8b03544

[30] Marin ML, Santos-Juanes L, Arques A, Amat AM, Miranda MA. Organic photocatalysts for the oxidation of pollutants and model compounds. Chemical Reviews. 2012;**112**:1710-1750. DOI: 10.1021/cr2000543

[31] Nguyen VH, Meejoo Smith S, Wantala K, Kajitvichyanukul P. Photocatalytic remediation of persistent organic pollutants (POPs): A review. Arabian Journal of Chemistry. 2020;**13**(11):8309-8337. DOI: 10.1016/j.arabjc.2020.04.028

[32] Long Z, Li Q, Wei T, Zhang G, Ren Z. Historical development and prospects of photocatalysts for pollutant removal in water. Journal of Hazardous Materials. 2020;**395**:122599. DOI: 10.1016/j.jhazmat.2020.122599

[33] Westerhoff P, Boyer T, Linden K. Emerging water technologies: Global pressures force innovation toward drinking water availability and quality. Accounts of Chemical Research. 2019;**52**:1146-1147. DOI: 10.1021/acs.accounts.9b00133

[34] Qu X, Brame J, Li Q, Alvarez PJJ. Nanotechnology for a safe and sustainable water supply: Enabling integrated water treatment and reuse. Accounts of Chemical Research. 2013;**46**(3):834-843. DOI: 10.1021/ar300029v

[35] Morin-Crini N, Lichtfouse E, Liu G, Crini G. Worldwide cases of water

pollution by emerging contaminants: A review. Environmental Chemistry Letters. 2022;**20**:2311-2338. DOI: 10.1007/s10311-022-01447-4

[36] Sauvé S, Desrosiers M. A review of what is an emerging contaminant. Chemistry Central Journal. 2014;**8**:15. DOI: 10.1186/1752-153X-8-15

[37] Fernandes JP, Almeida CMR, Salgado MA, Carvalho MF, Mucha AP. Pharmaceutical compounds in aquatic environments—Occurrence, fate and bioremediation prospective. Toxics. 2021;**9**(10):257. DOI: 10.3390/toxics9100257

[38] Ashiwaju B, Uzougbo C, Orikpete O. Environmental impact of pharmaceuticals: A comprehensive review. Matrix Science Pharma. 2024;**7**:85-94. DOI: 10.4103/mtsp.mtsp_15_23

[39] Taborda DA, Rodríguez BJ, Duque Agudelo BA. Disrupción endocrina en peces. Revista Colombiana de Ciencias Pecuarias. 2012;**25**(2):312-323

[40] Arukwe A. Cellular and molecular responses to endocrine modulators and the impact on fish reproduction. Marine Pollution Bulletin. 2001;**42**(8):643-655. DOI: 10.1016/s0025-326x(01)00062-5

[41] Samreen IA, Malak HA, Abulreesh HH. Environmental antimicrobial resistance and its drivers: A potential threat to public health. Journal of Global Antimicrobial Resistance. 2021;**27**:101-111. DOI: 10.1016/j.jgar.2021.08.001

[42] Correia A, Marcano L. Presencia y eliminación de compuestos farmacéuticos en plantas de tratamientos de aguas residuales: Revisión a nivel mundial y perspectiva nacional. Boletín de Malariología y Salud Ambiental. 2015;**55**(1):1-18

[43] Quesada Peñate I, Jáuregui Haza UJ, Wilhelm AM, Delmas H. Contaminación de las aguas con productos farmacéuticos. Estrategias para enfrentar la problemática. Revista CENIC Ciencias Biológicas. 2009;**40**(3):173-179. ISSN: 0253-5688

[44] Saravanathamizhan R, Perarasu VT. Chapter 6 - Improvement of biodegradability index of industrial wastewater using different pretreatment techniques. In: Shah MP, Sarkar A, Mandal S, editors. Wastewater Treatment. Elsevier; 2021. pp. 103-136. DOI: 10.1016/B978-0-12-821881-5.00006-4

[45] Silva JA. Wastewater treatment and reuse for sustainable water resources management: A systematic literature review. Sustainability. 2023;**15**(14):10940. DOI: 10.3390/su151410940

[46] Garrido-Cardenas JA, Esteban-García B, Agüera A, Sánchez-Pérez JA, Manzano-Agugliaro F. Wastewater treatment by advanced oxidation process and their worldwide research trends. International Journal of Environmental Research and Public Health. 2019;**17**(1):170. DOI: 10.3390/ijerph17010170

[47] Latifoglu A, Gürol MD. Advanced oxidation processes in water treatment. In: Hahn HH, Hoffmann E, Ødegaard H, editors. Chemical Water and Wastewater Treatment VI. Berlin, Heidelberg: Springer; 2000. DOI: 10.1007/978-3-642-59791-6_13

[48] Farrauto RJ, Dorazio L, Bartholomew CH. Introduction to Catalysis and Industrial Catalytic Processes. Reino Unido: Wiley; 2016

[49] Sad M, Castro M. Fundamentos de catálisis heterogénea. Santa Fe: Ediciones UNL; 2015. 197 p

[50] Upadhyay SK. Chemical Kinetics and Reaction Dynamics. Alemania: Anamaya Publishers; 2006

[51] Ramirez Sanabria AEMM. Catálisis heterogénea: principios básicos y su aplicación en química orgánica. Colombia: Universidad del Cauca; 2022

[52] Van Santen RA. Modern Heterogeneous Catalysis: An Introduction. Alemania: Wiley; 2017

[53] Viguera, Santiago E, Martínez BG. Materiales avanzados y nanomateriales: aprovechamiento de fuentes naturales y sus beneficios al medio ambiente. OmniaScience. 2022. pp. 181-184. ISBN 978-84-123480-3-3. DOI: 10.3926/oms.409

[54] Dutta K, Bhunia P, Vadivel S. Photocatalysts and Electrocatalysts in Water Remediation: From Fundamentals to Full Scale Applications. Reino Unido: Wiley; 2022. pp. 181-184. DOI: 10.3926/oms.409. Available from: https://www.omniascience.com/books/index.php/monographs/catalog/view/136/580/1099-1

[55] Yu J, Low J, Li X. Semiconductor Solar Photocatalysts: Fundamentals and Applications. Alemania: Wiley; 2021

[56] Kisch H. Semiconductor Photocatalysis: Principles and Applications. Alemania: Wiley; 2015

[57] Parrino F, Palmisano L. Titanium Dioxide (TiO2) and its Applications. Países Bajos: Elsevier Science; 2020

[58] Elsevier. Elsevier; 2001-2025. Available from: https://www.elsevier.com [Accessed: 19 September, 2024]

[59] American Chemical Society (ACS). American Chemical Society. 2001-2025. Available from: https://pubs.acs.org [Accessed: 19 September, 2024]

[60] SpringerLink. Springer; 2001-2025. Available from: https://link.springer.com [Accessed: 19 September, 2024]

[61] Khasawneh OFS, Palaniandy P. Occurrence and removal of pharmaceuticals in wastewater treatment plants. Process Safety and Environmental Protection. 2021;**150**:532-556. DOI: 10.1016/j.psep.2021.04.045

[62] Khasawneh OFS, Palaniandy P. Photocatalytic degradation of pharmaceuticals using TiO_2-based nanocomposite catalyst—Review. Civil and Environmental Engineering Reports. 2019;**29**(3):1-33. DOI: 10.2478/ceer-2019-0021

[63] Tong A, Braund R, Warren D, Peake B. TiO_2-assisted photodegradation of pharmaceuticals—A review. Open Chemistry. 2012;**10**(4):989-1027. DOI: 10.2478/s11532-012-0049-7

[64] Kanakaraju D, Glass BD, Oelgemöller M. Titanium dioxide photocatalysis for pharmaceutical wastewater treatment. Environmental Chemistry Letters. 2013;**12**(1):27-47. DOI: 10.1007/s10311-013-0428-0

[65] Friedmann D. A general overview of heterogeneous photocatalysis as a remediation technology for wastewaters containing pharmaceutical compounds. Water. 2022;**14**(21):3588. DOI: 10.3390/w14213588

[66] Ahmadpour N, Nowrouzi M, Avargani VM, Sayadi MH, Zendehboudi S. Design and optimization of TiO_2-based photocatalysts for efficient removal of pharmaceutical pollutants in water: Recent developments and challenges. Journal of Water Process Engineering. 2024;**57**:104597. DOI: 10.1016/j.jwpe.2023.104597

[67] Lee CM, Palaniandy P, Dahlan I. Pharmaceutical residues in aquatic environment and water remediation by TiO_2 heterogeneous photocatalysis: A review. Environmental Earth Sciences.

2017;**76**(17):611, 1-19. DOI: 10.1007/s12665-017-6924-y

[68] Kanakaraju D, Glass BD, Oelgemöller M. Advanced oxidation process-mediated removal of pharmaceuticals from water: A review. Journal of Environmental Management. 2018;**219**:189-207. DOI: 10.1016/j.jenvman.2018.04.103

[69] Sayadi MH, Chamanehpour E, Fahoul N. Recent advances and future outlook for treatment of pharmaceuticals from water: An overview. International Journal of Environmental Science and Technology. 2022;**20**(3):3437-3454. DOI: 10.1007/s13762-022-04674-y

[70] Aslam M, Fazal DB, Ahmad F, Fazal AB, Abdullah AZ, Ahmed M, et al. Photocatalytic degradation of recalcitrant pollutants of greywater. Catalysts. 2022;**12**(5):557. DOI: 10.3390/catal12050557

[71] Awofiranye OS, Modise SJ, Naidoo EB. Overview of polymer–TiO_2 catalyst for aqueous degradation of pharmaceuticals in heterogeneous photocatalytic process. Journal of Nanoparticle Research. 2020;**22**(6):168, 1-24. DOI: 10.1007/s11051-020-04877-9

[72] Sharma M, Yadav A, Mandal MK, Dubey KK. TiO_2 based photocatalysis: A valuable approach for the removal of pharmaceuticals from aquatic environment. International Journal of Environmental Science and Technology. 2022;**20**(4):4569-4584. DOI: 10.1007/s13762-021-03894-y

[73] Yang L, Yu LE, Ray MB. Degradation of paracetamol in aqueous solutions by TiO_2 photocatalysis. Water Research. 2008;**42**(13):3480-3488. DOI: 10.1016/j.watres.2008.04.023

[74] Foteinis S, Chatzisymeon E. Heterogeneous photocatalysis for water purification. In: Elsevier eBooks. 2020. pp. 75-97. DOI: 10.1016/b978-0-12-817836-2.00004-1

[75] Silva CG, Faria JL. Anatase vs. rutile efficiency on the photocatalytic degradation of clofibric acid under near UV to visible irradiation. Photochemical and Photobiological Sciences. 2009;**8**(5):705-711. DOI: 10.1039/b817364h

[76] Krakowiak R, Musial J, Bakun P, Spychała M, Czarczynska-Goslińska B, Mlynarczyk DT, et al. Titanium dioxide-based photocatalysts for degradation of emerging contaminants including pharmaceutical pollutants. Applied Sciences. 2021;**11**(18):8674. DOI: 10.3390/app11188674

[77] Gopinath KP, Madhav NV, Krishnan A, Malolan R, Rangarajan G. Present applications of titanium dioxide for the photocatalytic removal of pollutants from water: A review. Journal of Environmental Management. 2020;**270**:110906. DOI: 10.1016/j.jenvman.2020.110906

[78] Chen D, Cheng Y, Zhou N, Chen P, Wang Y, Li K, et al. Photocatalytic degradation of organic pollutants using TiO_2-based photocatalysts: A review. Journal of Cleaner Production. 2020;**268**:121725. DOI: 10.1016/j.jclepro.2020.121725

[79] Roslan NN, Lau HLH, Suhaimi NAA, Shahri NN, Verinda SB, Nur M, et al. Recent advances in advanced oxidation processes for degrading pharmaceuticals in wastewater—A review. Catalysts. 2024;**14**(3):189. DOI: 10.3390/catal14030189

[80] Nair N, Gandhi V, Shukla A, Ghotekar S, Nguyen V, Varma K. Mechanisms in the photocatalytic breakdown of persistent pharmaceutical

and pesticide molecules over TiO_2-based photocatalysts: A review. Journal of Physics: Condensed Matter. 2024;**36**(41):413003. DOI: 10.1088/1361-648x/ad5fd6

[81] Ojobe B, Zouzelka R, Satkova B, Vagnerova M, Nemeskalova A, Kuchar M, et al. Photocatalytic removal of pharmaceuticals from greywater. Catalysts. 2021;**11**(9):1125. DOI: 10.3390/catal11091125

[82] Beek TAD, Weber F, Bergmann A, Hickmann S, Ebert I, Hein A, et al. Pharmaceuticals in the environment—Global occurrences and perspectives. Environmental Toxicology and Chemistry. 2016;**35**(4):823-835. DOI: 10.1002/etc.3339

[83] Navidpour AH, Ahmed MB, Zhou JL. Photocatalytic degradation of pharmaceutical residues from water and sewage effluent using different TiO_2 nanomaterials. Nanomaterials. 2024;**14**(2):135. DOI: 10.3390/nano14020135

[84] Bayan E, Pustovaya L, Volkova M. Recent advances in TiO_2-based materials for photocatalytic degradation of antibiotics in aqueous systems. Environmental Technology and Innovation. 2021;**24**:101822. DOI: 10.1016/j.eti.2021.101822

[85] Pérez-Lucas G, Aatik AE, Aliste M, Navarro G, Fenoll J, Navarro S. Removal of contaminants of emerging concern from a wastewater effluent by solar-driven heterogeneous photocatalysis: A case study of pharmaceuticals. Water Air and Soil Pollution. 2023;**234**:55, 1-13. DOI: 10.1007/s11270-023-06075-4

[86] Shah AH, Rather MA. Pharmaceutical residues: New emerging contaminants and their mitigation by nano-photocatalysis. Advances in Nano Research. 2021;**10**(4):397-414. DOI: 10.12989/anr.2021.10.4.397

[87] Al-Nuaim MA, Alwasiti AA, Shnain ZY. The photocatalytic process in the treatment of polluted water. Chemical Papers. 2023;**77**(2):677-701. DOI: 10.1007/s11696-022-02468-7

[88] Akbari MZ, Xu Y, Lu Z, Peng L. Review of antibiotics treatment by advanced oxidation processes. Environmental Advances. 2021;**5**:100111. DOI: 10.1016/j.envadv.2021.100111

[89] Gatou MA, Syrrakou A, Lagopati N, Pavlatou EA. Photocatalytic TiO_2-based nanostructures as a promising material for diverse environmental applications: A review. Reactions. 2024;**5**(1):135-194. DOI: 10.3390/reactions5010007

[90] Tanji K, Navio JA, Chaqroune A, Naja J, Puga F, Hidalgo MC, et al. Fast photodegradation of rhodamine B and caffeine using ZnO-hydroxyapatite composites under UV-light illumination. Catalysis Today. 2022;**388**:176-186. DOI: 10.1016/j.cattod.2020.07.044

[91] Ijaz M, Zafar M. Titanium dioxide nanostructures as efficient photocatalyst: Progress, challenges and perspective. International Journal of Energy Research. 2021;**45**(3):3569-3589. DOI: 10.1002/er.6079

[92] Rengifo-Herrera JA, Pulgarin C. Why five decades of massive research on heterogeneous photocatalysis, especially on TiO_2, has not yet driven to water disinfection and detoxification applications? Critical review of drawbacks and challenges. Chemical Engineering Journal. 2023;**477**:146875. DOI: 10.1016/j.cej.2023.146875

[93] Peiris S, de Silva HB, Ranasinghe KN, Bandara SV, Perera IR. Recent development and future

prospects of TiO_2 photocatalysis. Journal of the Chinese Chemical Society. 2021;**68**(5):738-769. DOI: 10.1002/jccs.202000465

[94] Xochihua-Juan JL, Solis-Maldonado C, Luna-Sánchez RA, Enciso-Díaz OJ, Rojas-Ronquillo MR, Sandoval-Rangel L, et al. TiO_2 doped with europium (Eu): Synthesis, characterization and catalytic performance on pesticide degradation under solar irradiation. Catalysis Today. 2022;**394**:304-313. DOI: 10.1016/j.cattod.2021.08.024

[95] Navidpour AH, Abbasi S, Li D, Mojiri A, Zhou JL. Investigation of advanced oxidation process in the presence of TiO_2 semiconductor as photocatalyst: Property, principle, kinetic analysis, and photocatalytic activity. Catalysts. 2023;**13**(2):232. DOI: 10.3390/catal13020232

[96] Martín-Sómer M, Moreno-SanSegundo J, Álvarez-Fernández C, van Grieken R, Marugán J. High-performance low-cost solar collectors for water treatment fabricated with recycled materials, open-source hardware and 3D-printing technologies. Science of the Total Environment. 2021;**784**:147119. DOI: 10.1016/j.scitotenv.2021.147119

[97] Malato S, Maldonado MI, Fernandez-Ibanez P, Oller I, Polo I, Sánchez-Moreno R. Decontamination and disinfection of water by solar photocatalysis: The pilot plants of the Plataforma Solar de Almeria. Materials Science in Semiconductor Processing. 2016;**42**:15-23. DOI: 10.1016/j.mssp.2015.07.017

[98] Alvear-Daza JJ, Pais-Ospina D, Marín-Silva DA, Pinotti A, Damonte L, Pizzio LR, et al. Facile photocatalytic immobilization strategy for P-25 TiO_2 nanoparticles on low density polyethylene films and their UV-A photo-induced super hydrophilicity and photocatalytic activity. Catalysis Today. 2021;**372**:11-19. DOI: 10.1016/j.cattod.2020.07.029

[99] Joseph A, Vijayanandan A. Review on support materials used for immobilization of nano-photocatalysts for water treatment applications. Inorganica Chimica Acta. 2023;**545**:121284. DOI: 10.1016/j.ica.2022.121284

[100] Sivaraman C, Vijayalakshmi S, Leonard E, Sagadevan S, Jambulingam R. Current developments in the effective removal of environmental pollutants through photocatalytic degradation using nanomaterials. Catalysts. 2022;**12**(5):544. DOI: 10.3390/catal12050544

[101] Jiang D, Otitoju TA, Ouyang Y, Shoparwe NF, Wang S, Zhang A, et al. A review on metal ions modified TiO_2 for photocatalytic degradation of organic pollutants. Catalysts. 2021;**11**(9):1039. DOI: 10.3390/catal11091039

[102] Pujiastuti H, Kustiningsih I, Slamet S. Improvement of the efficiency of TiO_2 photocatalysts with natural dye sensitizers anthocyanin for the degradation of methylene blue. Jurnal Rekayasa Kimia and Lingkungan. 2021;**16**(2):84-99

[103] Ghalamchi L, Rasoulifard MH, Mohammadi Z, Dorraji MSS, Sehati N, Eskandarian MR. A solar-driven CPC photoreactor for decomposition of emerging contaminants in wastewater: Modeling and optimization. Chemical Engineering Research and Design. 2022;**182**:580-591. DOI: 10.1016/j.cherd.2022.04.032

[104] Díaz-Jiménez M, Sanjuan-Galindo R, Aba-Guevara C, Alonzo-García A, Mazón-Montijo DA,

Olivo-Alanis DS, et al. Effect of the implementation of static mixers in a CPC solar reactor for the photocatalytic degradation of paracetamol. Topics in Catalysis. 2022;**65**(9):980-988. DOI: 10.1007/s11244-022-01686-3

[105] Rapti I, Kosma C, Albanis T, Konstantinou I. Solar photocatalytic degradation of inherent pharmaceutical residues in real hospital WWTP effluents using titanium dioxide on a CPC pilot scale reactor. Catalysis Today. 2023;**423**:113884. DOI: 10.1016/j.cattod.2022.08.026

[106] Isopencu GO, Mocanu A, Deleanu IM. A brief review of photocatalytic reactors used for persistent pesticides degradation. ChemEngineering. 2022;**6**(6):89. DOI: 10.3390/chemengineering6060089

[107] Pérez JS, Arzate S, Soriano-Molina P, Sánchez JG, López JC, Plaza-Bolaños P. Neutral or acidic pH for the removal of contaminants of emerging concern in wastewater by solar photo-Fenton? A techno-economic assessment of continuous raceway pond reactors. Science of the Total Environment. 2020;**736**:139681. DOI: 10.1016/j.scitotenv.2020.139681

[108] Galushchinskiy A, González-Gómez R, McCarthy K, Farràs P, Savateev A. Progress in development of photocatalytic processes for synthesis of fuels and organic compounds under outdoor solar light. Energy and Fuels. 2022;**36**(9):4625-4639. DOI: 10.1021/acs.energyfuels.2c00178

[109] Rueda-Márquez JJ, Palacios-Villarreal C, Manzano M, Blanco E, del Solar MR, Levchuk I. Photocatalytic degradation of pharmaceutically active compounds (PhACs) in urban wastewater treatment plants effluents under controlled and natural solar irradiation using immobilized TiO_2. Solar Energy. 2020;**208**:480-492. DOI: 10.1016/j.solener.2020.08.028

[110] Zakria HS, Othman MHD, Kamaludin R, Kadir SHSA, Kurniawan TA, Jilani A. Immobilization techniques of a photocatalyst into and onto a polymer membrane for photocatalytic activity. RSC Advances. 2021;**11**(12):6985-7014. DOI: 10.1039/D0RA10964A

[111] Morshedy AS, Tawfik SM, Hashem KM, Abd El-Aty DM, Galhoum AA, Mostafa MS, et al. The production of clean diesel fuel by facile sun light photocatalytic desulfurization process using Cd-based diacetate as a novel liquid photocatalyst. Journal of Cleaner Production. 2021;**279**:123629

Chapter 9

The Quest for Sustainable Titanium Dioxide: Conventional and Sustainable Synthesis Approaches and Its Functionalities in Diverse Industries

Lorena Coelho, Mariana Ornelas, Bárbara R. Gomes and Bruna Moura

Abstract

In modern industry, interest in nanomaterials has grown significantly over the decades, and nanotechnology stands as a cutting-edge approach that has revolutionized industries worldwide. However, the current challenges related to nanomaterials synthesis revolve around replacing existing raw materials and processes with more sustainable alternatives while also addressing the need for scaling up production. One widely used compound is titanium dioxide (TiO_2), known for its chemical inertness, low cost, and high availability. It exhibits remarkable catalytic and distinctive semiconducting properties. This chapter will address the main physicochemical properties of TiO_2, which form the basis for its utilization across various fields, spanning a wide range of current and emerging applications. The case study will compare and discuss both conventional and sustainable preparation methods, considering material characterization and techniques to upscale production.

Keywords: TiO_2, $SiO_2@TiO_2$, core shell, upscaling, adsorption, photocatalysis, antibacterial

1. Introduction

Nanotechnology has rapidly advanced the development of a new generation of smart and innovative products and processes, creating tremendous growth potential for various sectors such as energy [1], health [2], environment [3], and agriculture [4]. The excitement around nanomaterials is driven by their unique and often superior properties, such as higher strength, lighter weight, better conductivity, and greater reactivity compared to their bulk counterparts. These properties are determined by factors such as the size, shape, surface area, composition, and structure of the nanomaterials.

The production of nanomaterials involves their synthesis, processing, characterization, and integration into larger systems and products. This process is characterized by a high degree of variability, uncertainty, and complexity, making it challenging to control, optimize, and scale up the quality and quantity of nanomaterials. Consequently, nanotechnology and the production of nanomaterials pose significant challenges, both technical and ethical, that need to be addressed before their full potential can be realized. Among these challenges are the complex and uncertain risks and benefits to human health and the environment, which are difficult to assess, predict, and manage [5], as well as the need to replace existing raw materials and processes with more sustainable alternatives. One factor hindering the scale-up of production is the scarcity and high cost of equipment specific to the processing of nanomaterials.

2. Titanium dioxide (TiO_2)

Titanium dioxide (TiO_2), an inorganic compound also known as titanium (IV) oxide, titanium white, or white pigment 6 in architectural paints and E171 in food additives, has been extensively studied and reported on by several authors in recent years [6]. As widely recognized, the use of materials in their nanoform offers distinctive and enhanced properties compared to the same material at the microscale due to their high surface area-to-volume ratio. TiO_2 nanoparticles possess unique properties that depend on factors such as particle size, distribution, and agglomeration [7].

TiO_2 finds widespread use in various applications due to its versatile properties, including chemical stability, optical characteristics, electrical behavior, thermal resistance, nontoxic nature, biocompatibility, photocatalytic properties, and low production cost [8–13]. The integration of TiO_2 is highly advantageous in pigments and dyes [14–16], since it is widely used as a white pigment in paints; plastics and cosmetics; air and water purification processes [13, 17–21], enhancing the removal of volatile organic pollutants, bacteria, viruses, and degradation of organic compounds; hydrogen production as an energy source [22, 23]; self-cleaning coatings [24]; photovoltaic cells [25, 26]; and batteries [27, 28].

The photocatalytic capacity of TiO_2 is its most important property due to its various technological and industrial applications, including air and water purification systems, soil remediation, and the sterilization of medical equipment and fabrics with antimicrobial properties [29–32]. The photocatalytic activity of TiO_2 is influenced by factors such as size, surface area, pore structure and volume, and crystalline phase [13]. In this context, the optimization of some of these properties has been the focus of the development of new forms of TiO_2.

2.1 Methods of synthesis

The main properties of TiO_2 can be fine-tuned during the synthesis of the materials. Obtaining different morphological forms of TiO_2 (i.e., nanoparticles, nanotubes, nanobuttons, and nanosheets), for example, is influenced by the synthesis process, depending on the type of precursors and solvents, pH, pressure and temperature, degree of crystallization, and duration of synthesis [33–36]. Conventional synthesis methods have some limitations, as they involve the use of hazardous chemical products and the production of toxic by-products that are not safe for the environment and also require high energy costs [37–40]. In this context, the approach of green, ecological, and sustainable methods is emerging to overcome the existing limitations.

The processes typically used for TiO_2 synthesis include the sol-gel method [33, 35, 36, 41], chemical precipitation [42, 43], hydrothermal [44–46], and chemical vapor deposition [47–49]. The sol-gel method is a promising alternative due to its simplicity, versatility, and low associated costs [33, 35–37]. Briefly, the synthesis of TiO_2 nanomaterials using the sol-gel method starts with the hydrolysis of a titanium precursor, followed by condensation and curing. It should be noted that the formation of Ti-O-Ti chains is favored by low water content and excess titanium alkoxide in the reaction [35]. This method also allows for stoichiometric control of the reaction, synthesis at low temperatures, incorporation of dopants, and the achievement of complex structures, making it suitable for large-scale production [33]. On the other hand, it is necessary to optimize certain conditions due to particle agglomeration. The use of surfactants can be an alternative to obtain nanoparticles with good size distribution and dispersion [35, 41]. Surfactants have a significant effect on the crystalline structure, uniformity, and distribution of the particles [35, 41, 50, 51].

The hydrothermal (HT) method is a more environmentally friendly alternative to conventional methods. This synthesis method begins with the formation of a homogeneous mixture by combining the precursor and the solvent [33, 44]. The mixture is then heated in a closed system at a specific pressure and temperature. The use of a closed reaction system allows for greater stability in the process, resulting in a high degree of crystallinity. HT processes are a promising alternative for the upscale production of nanomaterials, as they contribute to the sustainable production of nanomaterials, fuels, and materials from biomass [52]. HT processes use water as the reaction medium, eliminating the need for an energy-intensive drying step [53, 54]. As this route is aqueous based, the particles can be easily recovered from the reaction medium. It also improves the hydrophilicity of materials by increasing surface hydroxyl groups and allows access to photocatalytically active crystalline phases at much lower temperatures.

Nowadays, greener and more environmentally friendly methods of synthesizing TiO_2 are widely investigated, as they can overcome the limitations of conventional methods. In 2021, Aravind et al. reported the synthesis of TiO_2 nanoparticles (NPs) using jasmine flower extract as a reducing agent [38]. The color change from white to yellowish-gray confirmed the successful formation of TiO_2 NPs [38]. Another study explores the synthesis of TiO_2 NPs using different phytochemical sources obtained from various parts of plants (roots, leaves, flowers, and fruits) [55]. The secondary metabolites in the phytochemical extract are responsible for the conversion of metal ions into NPs [55, 56]. In this context, new greener synthesis methods for TiO_2 NPs are promising and are being explored, as they use natural sources and overcome some disadvantages of conventional methods.

2.2 Properties of TiO_2 nanomaterials

The widespread use of TiO_2 at nanoscale is due to its high surface area, permeability, and refractive index compared to conventional TiO_2 [57]. However, the use of these nanomaterials also raises concerns regarding toxicity and environmental impact, including ecotoxicity and bioaccumulation [8]. Therefore, the continuous optimization of TiO_2 properties, such as enhancing photocatalytic efficiency under visible light, ensuring medium- to long-term stability and durability, promoting sustainability, and addressing safety and environmental impact, is essential to meet future challenges and achieve the most suitable and favorable properties for TiO_2 applications.

TiO_2 exists in at least 11 different crystalline forms [12]. Anatase, rutile, and brookite are the most studied and widely used in various applications. **Figure 1** shows the crystal system of the main forms. The crystal structures correspond to Ti atoms surrounded by six O atoms in an octahedral configuration with small distortions [11, 59].

Anatase and rutile have a tetragonal crystal system, whereas brookite has an orthorhombic crystal system. Anatase contains chains of octahedrons that share edges in a single orientation, whereas rutile and brookite have two orientations [12]. The type of bonding in each crystal structure influences the properties of TiO_2. Anatase and rutile are the most commonly used polymorphic structures, as they have greater photocatalytic activity due to their highly ordered and oriented crystal structures [60]. Thermal stability increases in the order: anatase < brookite < rutile [8–12, 35, 60]. Anatase and brookite are metastable phases and can therefore undergo irreversible transformation to rutile at high temperatures [10, 11, 61]. The transformation of anatase to rutile occurs at temperatures between 400°C and 1200°C [60]. This range depends on the size of the crystallites and the presence of impurities. The transformation of brookite to rutile occurs at 800°C [60].

TiO_2 is also a semiconductor material, exhibiting a band gap of 3.0–3.2 eV [12]. The band gap of anatase is 3.20 eV (385 nm) and that of rutile is 3.02 eV (400 nm) [11, 59]. Only a small percentage of the solar spectrum (<5%), corresponding to UV light, can be utilized by TiO_2 in photocatalytic processes. However, this limitation can be mitigated by incorporating other elements and/or interacting with other photocatalytic materials [35, 61]. Indeed, the incorporation of different elements into TiO_2 allows for better performance and efficiency of the material, as reported by Jang and Yu, who found that the SiO_2@ material is more efficient for the photocatalytic decomposition of Rhodamine–6G due to the adsorption capacity of SiO_2 [62, 63]. However, TiO_2 doping introduces impurities into the material, which causes a decrease in crystallinity and, consequently, a decrease in the efficiency of the photocatalytic activity [64].

Several studies have been carried out to optimize the properties of TiO_2. The approach is to dope TiO_2 with different elements (metals and nonmetals) in order to increase its photocatalytic activity by creating new energy levels near the conduction band [65–68]. The production of black TiO_2 is a promising approach for future needs,

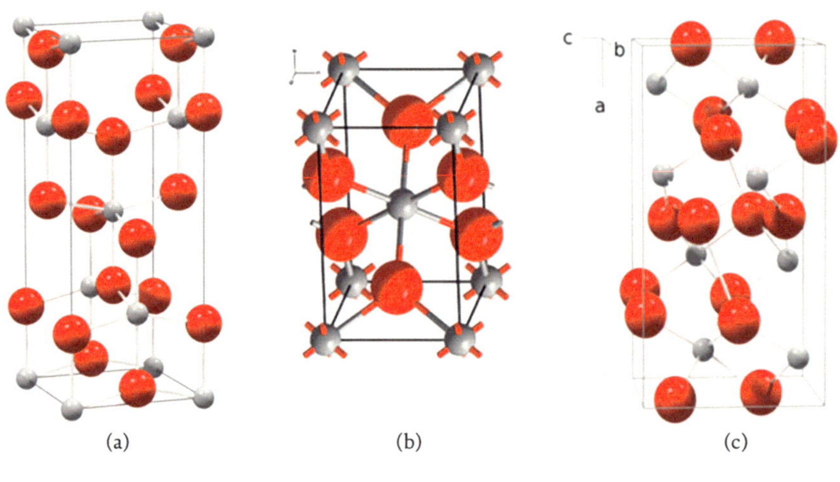

(a) (b) (c)

Figure 1.
Crystal system of (a) rutile, (b) anatase, and (c) brookite TiO_2 [58].

as its optical properties have been highly optimized. For example, Chen et al. first reported the production of black TiO_2 by exposing white TiO_2 to a temperature of 200°C under 20 bar H_2 pressure for 5 days [69]. These conditions led to the formation of defects with high visible light absorption [69, 70].

2.3 Upscaling the production of TiO_2 nanomaterials

The upscaling of nanomaterials, particularly TiO_2, has been a focal point of research due to its broad applications [71–77]. In 2024, for example, Wu et al. reported a highly reproducible upscaled synthesis of 200–800 nm amorphous and anatase TiO_2 NPs based on a combination of the sol-gel method with a hydrothermal process, performed on a 100-mL scale [77].

Nevertheless, the transition from laboratory-scale synthesis to industrial-scale production presents significant challenges, and the commonly used fabrication techniques for TiO_2 nanomaterials, such as sol-gel processes, hydrothermal methods, and chemical vapor deposition, face issues related to cost, uniformity, and reproducibility when scaled up [78]. Moreover, the control of nanoparticle size, morphology, and crystalline phase becomes increasingly difficult in large-scale production, which can significantly impact the performance of the final product [78]. Innovative approaches and continuous advancements in synthesis techniques are crucial to overcoming these obstacles, ensuring that the high-performance characteristics of TiO_2 observed at the laboratory scale can be maintained in industrial applications.

3. Case of study

The project *NanoTheC-Aba – CECs and AMR bacteria pre-concentration by ultra-nano filtration and Abatement by ThermoCatalytic Nano-powders implementing circular economy solution* is a European project aimed at developing an integrated system that provides abatement of a wide spectrum of contaminants of emerging concern (CEC) and antimicrobial resistant (AMR) pathogens, along with the complete reuse of process effluents, thereby minimizing the disposal of wastewater into the environment [79].

The project will deliver an energy-efficient, new integrated prototype system for water purification, composed of the first-to-be-realized ultra-stable silicon carbide (SiC) ultrafiltration/nanofiltration (UF/NF) membrane, an innovative nano-enabled thermocatalytic energy-efficient packed-bed reactor (TPBR), and a nano-enabled antimicrobial microfiltration (MF) membrane (**Figure 2**).

One of the components of the integrated system is a microfiltration prefiltering unit with antibacterial functionality (AM-MF), achieved by using antimicrobial nanomaterials immobilized on an SiC flat membrane.

In line with circular economy principles, SiO_2@TiO_2 core-shell nanomaterial was developed using sustainable synthesis strategies. The viability of scaling up the synthesis of these materials was assessed by comparing them with those produced through laboratory-scale synthesis [21].

As a case study, results from the characterization of nanomaterials developed and reported in Ref. [21] will be presented and discussed. Conventional pollution remediation methods have limited effectiveness. For example, water treatment techniques like adsorption or coagulation only transfer pollutants to other phases, while sedimentation, filtration, chemical, and membrane technologies incur high costs and can produce toxic secondary pollutants [80]. Consequently, scientific interest has shifted

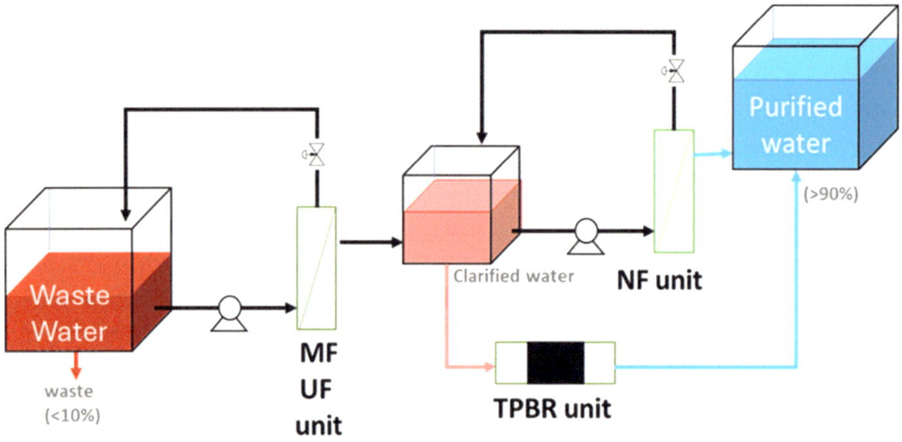

Figure 2.
Illustrative scheme of the NanoTheC-Aba prototype system for water purification.

toward alternative methods, such as advanced oxidation processes (AOPs), which include strategies like semiconductor-based photocatalysis. These methods offer innovative alternatives by aiming to fully mineralize organic pollutants into harmless by-products like O_2, H_2O, N_2, and mineral acids [81, 82].

The synthesis of TiO_2 nanoparticles plays a crucial role in the production of wastewater treatment membranes. These nanoparticles exhibit excellent photocatalytic properties, enabling the degradation of organic pollutants and enhancing overall treatment efficiency. Scaling up their production ensures wider application and more effective water purification. Therefore, as part of the NanoTheC-Aba project, one of the main objectives of this work was to assess the viability of scaling up the synthesis of SiO_2@TiO_2 core-shell nanomaterials by comparing the obtained particles with those produced using laboratory-scale synthesis. To this purpose, the hydrothermal (HT) method was chosen to obtain the core-shell material, and SiO_2 and TiO_2 P25 (commercial) were used as comparative and reference materials. The improved properties of SiO_2@TiO_2 core-shell NPs result from the synergy between the photocatalytic and self-cleaning properties of TiO_2 and the adsorptive and mechanical properties of SiO_2. As a result, enhanced performance in both photocatalysis and adsorption is anticipated.

The SiO_2@TiO_2 core-shell synthesis process can be divided into the following steps: first, the monodisperse SiO_2 cores are prepared by the classical Stöber method [83]; then the synthesized SiO_2 core is coated with a thin layer of TiO_2, followed by calcination (Calc., 650°C for 2 h) or HT treatments (140°C for 6 h) [84]. Although the upscale process has several reported limitations, a tenfold upscaling of the SiO_2@TiO_2 core-shell NPs was achieved in this work. The physicochemical properties, as well as the performance of the obtained materials, were assessed.

3.1 TiO_2 physicochemical characterization

The morphology was analyzed using high-resolution transmission electron microscopy (HR-TEM) and field emission scanning electron microscopy (FE-SEM). **Figure 3** shows the spherical, well-formed particles of SiO_2 with a smooth surface and

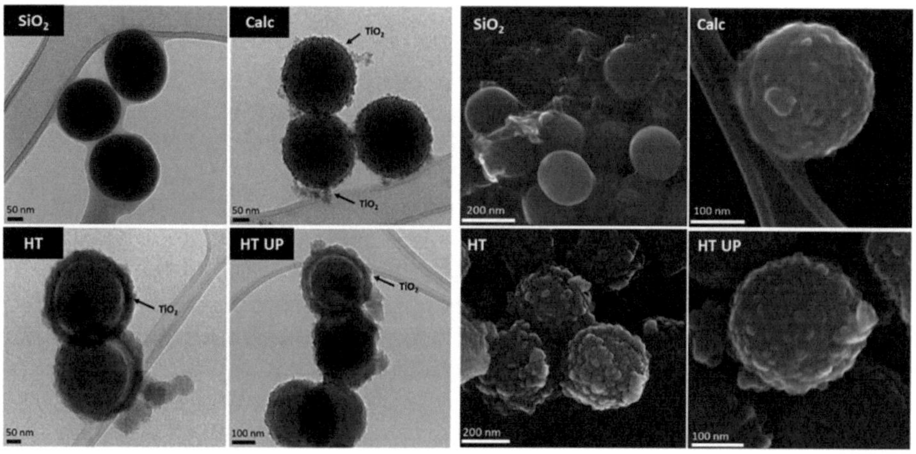

Figure 3.
HR-TEM (left) and FE-SEM (right) of SiO_2 core, $SiO_2@TiO_2$ calcined core-shell NPs, $SiO_2@TiO_2$ core-shell NPs HT, and $SiO_2@TiO_2$ core-shell NPs HT upscaled.

no evident defects. In contrast, the synthesized $SiO_2@TiO_2$ core-shell nanoparticles (NPs), both calcined and hydrothermally treated (HT), exhibit a rough and textured surface, suggesting that the TiO_2 was successfully coated on the silica particles. Furthermore, it is evident from the images that the shell formed during the HT treatment is larger than that formed by the calcination process.

Additionally, the morphology of the upscaled NPs is similar to that obtained at the laboratory scale. These results suggest that the upscaling process was successful, demonstrating its feasibility and reproducibility. Upscaling is known to be one of the main challenges in nanoparticle synthesis development.

As presented in **Table 1**, the size of the core-shell NPs increases compared to the SiO_2 core, due to the formation of the TiO_2 layer. The $SiO_2@TiO_2$ NPs prepared by calcination method show a slightly increased average diameter compared with those produced by the hydrothermal process, and as evidenced by the morphological analysis (**Figure 3**).

The specific surface area (SSA), determined by N_2 gas-volumetric adsorption at 77 K, increases for $SiO_2@TiO_2$, in both thermal treatments, suggesting the presence of nanocrystalline TiO_2 particles, since they have a larger surface area due to their small size [84]. A higher increase in SSA was observed for the HT treatment, indicating that

	Size (nm)	BET SSA (m^2/g)
SiO_2	185 ± 11	18
TiO_2 P25	21 (datasheet)	52
$SiO_2@TiO_2$ Calc.	211 ± 33	52
$SiO_2@TiO_2$ HT	333 ± 30	304
$SiO_2@TiO_2$ HT UP	284 ± 24	280

Table 1.
Nanomaterials size and specific surface area (SSA) obtained by BET analysis.

this method contributes to the formation of smaller TiO_2 particles in the shell of the composite materials, resulting in a greater SSA.

The chemical composition analysis, obtained by attenuated total reflectance Fourier transform infrared (FT-IR ATR) spectroscopy, is presented in **Figure 4**. In the spectra, symmetric and asymmetric stretching modes of the Si-O-Si bond are observed around 790 and 1110 cm^{-1} [85]. The intensity of the Si-O-Si bond is higher in $SiO_2@TiO_2$ samples that have been calcined compared to the HT samples. This difference can be attributed to the thinner TiO_2 shell in the calcined sample, as opposed to the thicker shell in HT samples.

The band around 1400 cm^{-1} is attributed to Ti–O–Ti vibration and is observed for both TiO_2 and all $SiO_2@TiO_2$ samples. The intensity of this band is higher for the HT samples compared to the calcined and reference samples, which can be related to the thickness of the TiO_2 shell. The band around 970 cm^{-1} is attributed to the Ti-O-Si bond structure and is present in all core-shell NPs [86], confirming the formation of the titanium shell bonded to the silica core [87].

The spectrum of the $SiO_2@TiO_2$ (HT) upscaled sample presents identical peaks to those of the sample synthesized at the laboratory scale, suggesting that scaling up the production process does not affect the chemical composition of the NPs.

The crystalline structures and phases were confirmed by X-ray diffraction (XRD) analysis (**Figure 5**). The synthesized SiO_2 and $SiO_2@TiO_2$ NPs exhibit broad XRD peaks typical of amorphous and/or nanometric powders, while commercial TiO_2 is a fully crystalline powder composed of anatase and rutile phases [88, 89]. Although the $SiO_2@TiO_2$ core-shell NPs show lower crystallinity compared to the commercial TiO_2, the main characteristic peaks of the anatase (101, 200) and rutile (110) phases are still observable.

3.2 TiO_2 performance evaluation

The dye removal capacity, relevant for wastewater treatment applications, was evaluated for the developed nanomaterial. The adsorption capacity and photocatalytic

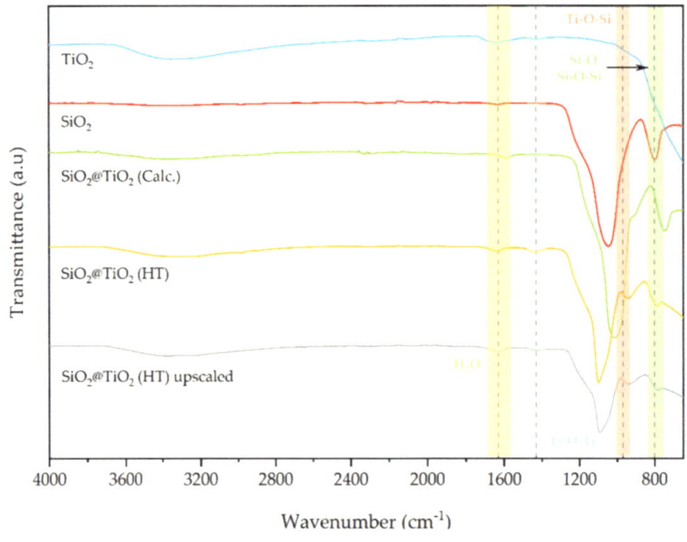

Figure 4.
Normalized FT-IR ATR spectra of the synthesized $SiO_2@TiO_2$ NPs compared with pure SiO_2 and TiO_2.

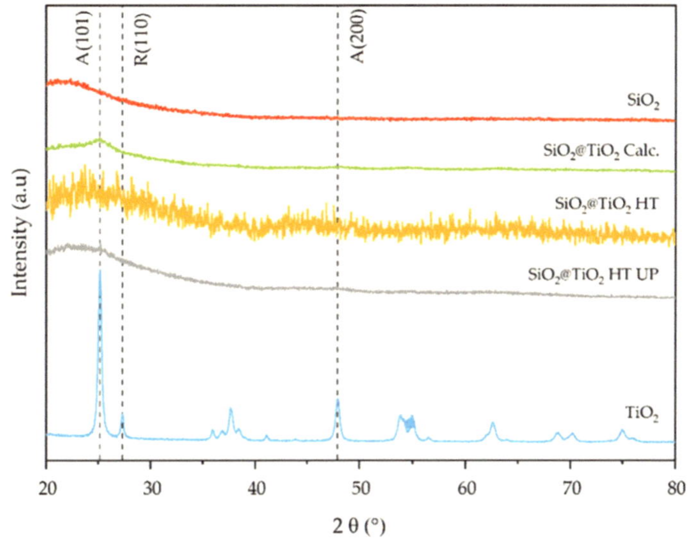

Figure 5.
X-ray diffraction patterns of synthesized $SiO_2@TiO_2$ NPs, compared with pure SiO_2 and TiO_2.

properties of the synthesized nanomaterial were assessed using methylene blue (MB) in an aqueous solution as a model dye. The removal efficiency of the nanomaterials was examined through their ability to remove MB under UV light irradiation. The obtained results are presented in **Figure 6**. After UV light exposure, the total concentration of MB was determined from the maximum absorption measurements using UV/Vis spectroscopy, with the characteristic peak of MB dye at 663 nm being used to study its catalytic degradation [90].

The SiO_2 and $SiO_2@TiO_2$ NPs presented a remarkable adsorption capacity compared to commercial TiO_2. This effect was even more pronounced for the HT samples, which can be attributed to their higher SSA compared to the calcined and commercial samples. Before UV exposure, under dark conditions, the HT samples almost

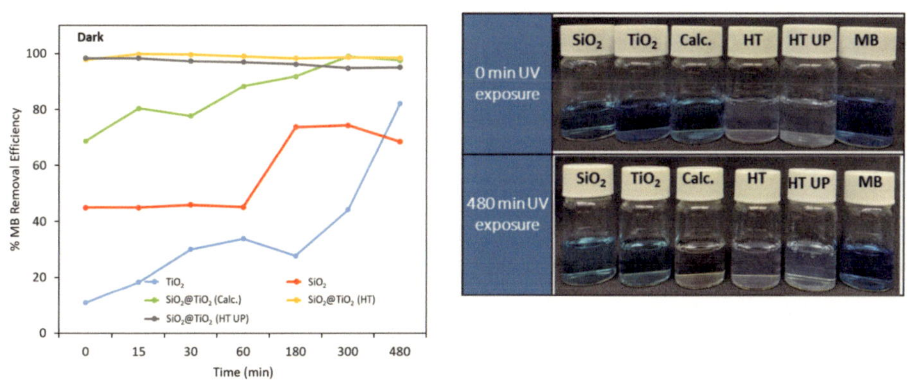

Figure 6.
Methylene blue removal efficiency (left) and images of the vials containing the MB solution before (0 min) and after 480 min of UV exposure with the powdered samples (right).

completely degraded the MB dye. After UV exposure, the core-shell NPs demonstrated higher MB removal and outperformed commercial TiO_2 in MB removal. This fact illustrated the synergistic effect of both adsorption and photocatalysis.

Although TiO_2 is the most commonly used semiconductor material for photocatalysis due to its chemical stability [91], this work demonstrates that an SiO_2@TiO_2 core-shell structure improves dye degradation by leveraging the adsorption capacity of SiO_2.

Following the successful upscaling of the nanomaterial, which exhibited physicochemical properties and performance similar to those observed in the laboratory batch, the antibacterial activity was evaluated under dynamic contact and dark conditions according to the standard ASTM E2149:2020. The results showed a reduction of 97.4% for *Staphylococcus aureus* and of 92.7% for *Escherichia coli*.

4. Conclusions

TiO_2, a versatile nanomaterial renowned for its photocatalytic, antibacterial, and adsorption properties, stands out due to its high stability, nontoxicity, and cost-effectiveness, making it an ideal candidate for various industrial applications.

The sustainable synthesis of TiO_2 promises to reduce environmental impact while maintaining the material's superior properties. However, upscaling the production of TiO_2 from laboratory to industrial scale presents significant challenges. These include maintaining the uniformity and reproducibility of nanoparticle size, morphology, and crystalline phase, as well as managing the high costs and complexities associated with large-scale manufacturing processes.

When it comes to producing SiO_2@TiO_2 core-shell nanoparticles, the choice between calcination and HT treatment significantly impacts environmental sustainability. Although the particles exhibit excellent photocatalytic properties and adsorption capacity regardless of the production method, the HT method is more energy efficient. HT typically uses lower temperatures, thus reducing the energy consumption associated with the calcination process. The HT method has proven to be a promising, sustainable alternative for the upscaled synthesis of nanomaterials, ensuring reproducibility of production.

In conclusion, the unique properties and diverse applications of TiO_2 underscore its significant potential in driving technological innovations and addressing global challenges. Future research and development in sustainable synthesis methods, as well as overcoming upscaling challenges, will be crucial for unlocking the full potential of TiO_2 in various fields.

Acknowledgements

This project has received funding from the European Union's Horizon 2020 research and innovation programme under grant agreement No. 869178-AquaticPollutants.

Conflict of interest

The authors declare no conflict of interest.

Thanks

Giuliana Magnacca, from Dipartimento di Chimica and NIS Interdepartmental Centre from Università degli Studi di Torino (UNITO), and Francesca Deganello, from Consiglio Nazionale delle Ricerche (CNR) Istituto per lo Studio dei Materiali Nanostrutturati (ISMN), are greatly acknowledged for their support in some of the measurements.

Author details

Lorena Coelho*, Mariana Ornelas, Bárbara R. Gomes and Bruna Moura
CeNTI—Centre for Nanotechnology and Advanced Materials,
Vila Nova de Famalicão, Portugal

*Address all correspondence to: lcoelho@centi.pt

References

[1] Ahmadi MH, Ghazvini M, Alhuyi Nazari M, Ahmadi MA, Pourfayaz F, Lorenzini G, et al. Renewable energy harvesting with the application of nanotechnology: A review. International Journal of Energy Research. 2019;**43**(4):1387-1410

[2] Anselmo AC, Mitragotri S. Nanoparticles in the clinic: An update. Bioengineering & Translational Medicine. 2019;**4**(3):e10143

[3] Usman M, Farooq M, Wakeel A, Nawaz A, Cheema SA, Rehman H u, et al. Nanotechnology in agriculture: Current status, challenges and future opportunities. Science of the Total Environment. 2020;**721**:137778

[4] Prasad R, Bhattacharyya A, Nguyen QD. Nanotechnology in sustainable agriculture: Recent developments, challenges, and perspectives. Frontiers in Microbiology. 2017;**8**. [cited 2024 May 27]. Available from: https://www.frontiersin.org/journals/microbiology/articles/10.3389/fmicb.2017.01014/full

[5] Najahi-Missaoui W, Arnold RD, Cummings BS. Safe nanoparticles: Are we there yet? International Journal of Molecular Sciences. 2021;**22**(1):385

[6] Lan Y, Lu Y, Ren Z. Mini review on photocatalysis of titanium dioxide nanoparticles and their solar applications. Nano Energy. 2013;**2**(5):1031-1045

[7] Hsu CY, Mahmoud ZH, Abdullaev S, Ali FK, Ali Naeem Y, Mzahim Mizher R, et al. Nano titanium oxide (nano-TiO_2): A review of synthesis methods, properties, and applications. Case Studies in Chemical and Environmental Engineering. 2024;**9**:100626

[8] Rashid MM, Forte Tavčer P, Tomšič B. Influence of titanium dioxide nanoparticles on human health and the environment. Nanomaterials. 2021;**11**(9):2354

[9] Ali I, Suhail M, Alothman ZA, Alwarthan A. Recent advances in syntheses, properties and applications of TiO_2 nanostructures. RSC Advances. 2018;**8**(53):30125-30147

[10] Mhadhbi M, Abderazzak H, Avar B, Mhadhbi M, Abderazzak H, Avar B. Synthesis and properties of titanium dioxide nanoparticles. In: Updates on Titanium Dioxide. London, UK: IntechOpen; 2023. [cited 2024 May 17]. Available from: https://www.intechopen.com/chapters/87015

[11] Chen X, Selloni A. Introduction: Titanium dioxide (TiO_2) nanomaterials. Chemical Reviews. 2014;**114**(19):9281-9282

[12] Parrino F, Palmisano L. Titanium Dioxide (TiO_2) and Its Applications. 1st edition. Amsterdam: Elsevier; 2020. 735 p. ISBN 9780128199602

[13] Nakata K, Fujishima A. TiO_2 photocatalysis: Design and applications. Journal of Photochemistry and Photobiology C Photochemistry Review. 2012;**13**(3):169-189

[14] Goulart S, Jaramillo Nieves LJ, Dal Bó AG, Bernardin AM. Sensitization of TiO_2 nanoparticles with natural dyes extracts for photocatalytic activity under visible light. Dyes and Pigments. 2020;**182**:108654

[15] Braun JH, Baidins A, Marganski RE. TiO_2 pigment technology: A review. Progress in Organic Coating. 1992;**20**(2):105-138

[16] Andrzejewska A, Krysztafkiewicz A, Jesionowski T. Adsorption of organic dyes on the aminosilane modified TiO_2 surface. Dyes and Pigments. 2004;**62**(2):121-130

[17] Patil SB, Basavarajappa PS, Ganganagappa N, Jyothi MS, Raghu AV, Reddy KR. Recent advances in non-metals-doped TiO_2 nanostructured photocatalysts for visible-light driven hydrogen production, CO_2 reduction and air purification. International Journal of Hydrogen Energy. 2019;**44**(26):13022-13039

[18] Ao CH, Lee SC. Indoor air purification by photocatalyst TiO_2 immobilized on an activated carbon filter installed in an air cleaner. Chemical Engineering Science. 2005;**60**(1):103-109

[19] Mohammed SS, Shnain ZY, Abid MF. Use of TiO_2 in photocatalysis for air purification and wastewater treatment: A review. Engineering and Technology Journal. 2022;**40**(9):1131-1143

[20] Perović K, dela Rosa FM, Kovačić M, Kušić H, Štangar UL, Fresno F, et al. Recent achievements in development of TiO_2-based composite photocatalytic materials for solar driven water purification and water splitting. Materials. 2020;**13**(6):1338

[21] Gomes BR, Lopes JL, Coelho L, Ligonzo M, Rigoletto M, Magnacca G, et al. Development and upscaling of SiO_2@TiO_2 core-shell nanoparticles for methylene blue removal. Nanomaterials. 2023;**13**(16):2276

[22] Miyoshi A, Nishioka S, Maeda K. Water splitting on rutile TiO_2-based photocatalysts. Chemistry—A European Journal. 2018;**24**(69):18204-18219

[23] Eidsvåg H, Bentouba S, Vajeeston P, Yohi S, Velauthapillai D. TiO_2 as a photocatalyst for water splitting—An experimental and theoretical review. Molecules. 2021;**26**(6):1687

[24] Lukong VT, Ukoba K, Jen TC. Review of self-cleaning TiO_2 thin films deposited with spin coating. International Journal of Advanced Manufacturing Technology. 2022;**122**(9):3525-3546

[25] Nogueira AF, De Paoli MA. A dye sensitized TiO_2 photovoltaic cell constructed with an elastomeric electrolyte. Solar Energy Materials & Solar Cells. 2000;**61**(2):135-141

[26] Sinha D, De D, Goswami D, Mondal A, Ayaz A. ZnO and TiO_2 nanostructured dye sensitized solar photovoltaic cell. Materials Today Proceedings. 2019;**11**:782-788

[27] Fehse M, Ventosa E. Is TiO_2(B) the future of titanium-based battery materials? ChemPlusChem. 2015;**80**(5):785-795

[28] Usui H, Yoshioka S, Wasada K, Shimizu M, Sakaguchi H. Nb-doped rutile TiO_2: A potential anode material for Na-ion battery. ACS Applied Materials & Interfaces. 2015;**7**(12):6567-6573

[29] Chung CJ, Lin HI, Tsou HK, Shi ZY, He JL. An antimicrobial TiO_2 coating for reducing hospital-acquired infection. Journal of Biomedical Materials Research. Part B, Applied Biomaterials. 2008;**85B**(1):220-224

[30] Wang A n, Teng Y, Hu X f, Wu L h, Huang Y j, Luo Y m, et al. Diphenylarsinic acid contaminated soil remediation by titanium dioxide (P25) photocatalysis: Degradation pathway, optimization of operating parameters and effects of soil properties. Science of the Total Environment. 2016;**541**:348-355

[31] Higarashi MM, Jardim WF. Remediation of pesticide contaminated soil using TiO2 mediated by solar light. Catalysis Today. 2002;**76**(2):201-207

[32] Luo L, Miao L, Tanemura S, Tanemura M. Photocatalytic sterilization of TiO_2 films coated on Al fiber. Materials Science and Engineering B. 2008;**148**(1):183-186

[33] Kumarage GWC, Hakkoum H, Comini E. Recent advancements in TiO_2 nanostructures: Sustainable synthesis and gas sensing. Nanomaterials. 2023;**13**(8):1424

[34] Suhan MBK, Al-Mamun MR, Farzana N, Aishee SM, Islam MS, Marwani HM, et al. Sustainable pollutant removal and wastewater remediation using TiO_2-based nanocomposites: A critical review. Nano-Structures & Nano-Objects. 2023;**36**:101050

[35] Chen X, Mao SS. Titanium dioxide nanomaterials: Synthesis, properties, modifications, and applications. Chemical Reviews. 2007;**107**(7):2891-2959

[36] Raghavan S. TiO_2 nanostructures by sol-gel processing. In: Sol-Gel Method—Recent Advances. London, UK: IntechOpen; 2023. [cited 2024 Jun 28]. Available from: https://www.intechopen.com/chapters/87067

[37] Eddy DR, Rahmawati D, Permana MD, Takei T, Solihudin S, et al. A review of recent developments in green synthesis of TiO_2 nanoparticles using plant extract: Synthesis, characterization and photocatalytic activity. Inorganic Chemistry Communications. 2024;**165**:112531

[38] Aravind M, Amalanathan M, Mary MSM. Synthesis of TiO_2 nanoparticles by chemical and green synthesis methods and their multifaceted properties. SN Applied Sciences. 2021;**3**(4):409

[39] Rana A, Yadav K, Jagadevan S. A comprehensive review on green synthesis of nature-inspired metal nanoparticles: Mechanism, application and toxicity. Journal of Cleaner Production. 2020;**272**:122880

[40] Irshad MA, Nawaz R, Zia ur Rehman M, Imran M, Ahmad J, Ahmad S, et al. Synthesis and characterization of titanium dioxide nanoparticles by chemical and green methods and their antifungal activities against wheat rust. Chemosphere. 2020;**258**:127352

[41] Lukong VT, Ukoba KO, Jen TC. Heat-assisted sol–gel synthesis of TiO_2 nanoparticles structural, morphological and optical analysis for self-cleaning application. Journal of King Saud University—Science. 2022;**34**(1):101746

[42] Wang Z, Liu S, Cao X, Wu S, Liu C, Li G, et al. Preparation and characterization of TiO_2 nanoparticles by two different precipitation methods. Ceramics International. 2020;**46**(10, Part A):15333-15341

[43] Buraso W, Lachom V, Siriya P, Laokul P. Synthesis of TiO_2 nanoparticles via a simple precipitation method and photocatalytic performance. Materials Research Express. 2018;**5**(11):115003

[44] Waseem S, Zeeshan T, Tariq H, Majid F, Ali MD, Kayani ZN, et al. The influence of transition metals (Fe, Co) on the structural, magnetic and optical properties of TiO_2 nanoparticles synthesized by the hydrothermal method. Applied Physics A: Materials Science & Processing. 2022;**128**(8):690

[45] Gomathi Thanga Keerthana B, Solaiyammal T, Muniyappan S,

Murugakoothan P. Hydrothermal synthesis and characterization of TiO_2 nanostructures prepared using different solvents. Materials Letters. 2018;**220**:20-23

[46] Gupta T, Samriti CJ, Prakash J. Hydrothermal synthesis of TiO_2 nanorods: Formation chemistry, growth mechanism, and tailoring of surface properties for photocatalytic activities. Materials Today Chemistry. 2021;**20**:100428

[47] Liu W, Tang H, Liu D. Combining density functional theory and CFD-PBM model to predict TiO_2 nanoparticle evolution during chemical vapor deposition. Chemical Engineering Journal. 2023;**454**:140174

[48] Aghaee M, Verheyen J, Stevens AAE, Kessels WMM, Creatore M. TiO_2 thin film patterns prepared by chemical vapor deposition and atomic layer deposition using an atmospheric pressure microplasma printer. Plasma Processes and Polymers. 2019;**16**(12):1900127

[49] Wang WB, Yanguas-Gil A, Yang Y, Kim DY, Girolami GS, Abelson JR. Chemical vapor deposition of TiO_2 thin films from a new halogen-free precursor. Journal of Vacuum Science and Technology A. 2014;**32**(6):061502

[50] Pierre AC. Surfactant-Templated sol-gel materials. In: Pierre AC, editor. Introduction to sol-Gel Processing [Internet]. Cham: Springer International Publishing; 2020. [cited 2024 Jun 28]. Available from:. pp. 457-495. DOI: 10.1007/978-3-030-38144-8_11

[51] Masoudipour E, Kashanian S, Azandaryani AH, Omidfar K, Bazyar E. Surfactant effects on the particle size, zeta potential, and stability of starch nanoparticles and their use in a pH-responsive manner. Cellulose. 2017;**24**(10):4217-4234

[52] Zhang B, Biswal BK, Zhang J, Balasubramanian R. Hydrothermal treatment of biomass feedstocks for sustainable production of chemicals, fuels, and materials: Progress and perspectives. Chemical Reviews. 2023;**123**(11):7193-7294

[53] Gao Y, Remón J, Matharu AS. Microwave-assisted hydrothermal treatments for biomass valorisation: A critical review. Green Chemistry. 2021;**23**(10):3502-3525

[54] Mamaghani AH, Haghighat F, Lee CS. Hydrothermal/solvothermal synthesis and treatment of TiO_2 for photocatalytic degradation of air pollutants: Preparation, characterization, properties, and performance. Chemosphere. 2019;**219**:804-825

[55] Roy J. The synthesis and applications of TiO_2 nanoparticles derived from phytochemical sources. Journal of Industrial and Engineering Chemistry. 2022;**106**:1-19

[56] Ajitha B, Reddy YAK, Reddy PS. Biosynthesis of silver nanoparticles using *Momordica charantia* leaf broth: Evaluation of their innate antimicrobial and catalytic activities. Journal of Photochemistry and Photobiology. B. 2015;**146**:1-9

[57] Kusior A, Banas J, Trenczek-Zajac A, Zubrzycka P, Micek-Ilnicka A, Radecka M. Structural properties of TiO_2 nanomaterials. Journal of Molecular Structure. 2018;**1157**:327-336

[58] Samat MH, Ali AMM, Taib MFM, Hassan OH, Yahya MZA. Hubbard U calculations on optical properties of 3*d* transition metal oxide TiO_2. Results in Physics. 2016;**6**:891-896

[59] Baccaro ALB, Cordon LD, Nishimura FG, Gutz IGR. Fotocatálise

Mediada Por Tio_2 No Estado Nanoparticulado: Revisão Da Reatividade Pelo Conceito De *Traps* E Algumas Aplicações Em Química Analítica. Química Nova. 2019;**42**:329-345

[60] Boytsova O, Zhukova I, Tatarenko A, Shatalova T, Beiltiukov A, Eliseev A, et al. The anatase-to-rutile phase transition in highly oriented nanoparticles array of titania with photocatalytic response changes. Nanomaterials. 2022;**12**(24):4418

[61] Ahmadlouydarab M, Javadi S, Darab AA, F. Evaluation of thermal stability of TiO_2 applied on the surface of a ceramic tile to eliminate methylene blue using silica-based doping materials. Advanced Journal of Chemistry, Section A. 2023;**6**(4):352-365

[62] Anderson C, Bard AJ. Improved photocatalytic activity and characterization of mixed TiO_2/SiO_2 and TiO_2/Al_2O_3 materials. The Journal of Physical Chemistry. B. 1997;**101**(14):2611-2616

[63] Jang M, Yu Z. Modeling heterogeneous oxidation of NOx, SO_2 and hydrocarbons in the presence of mineral dust particles under various atmospheric environments. In: Multiphase Environmental Chemistry in the Atmosphere [Internet]. American Chemical Society; 2018. pp. 301-326. (ACS Symposium Series; vol. 1299). [cited 2024 May 31]. Available from:. DOI: 10.1021/bk-2018-1299.ch015

[64] Ullattil SG, Narendranath SB, Pillai SC, Periyat P. Black TiO_2 nanomaterials: A review of recent advances. Chemical Engineering Journal. 2018;**343**:708-736

[65] Khlyustova A, Sirotkin N, Kusova T, Kraev A, Titov V, Agafonov A. Doped TiO_2: The effect of doping elements on photocatalytic activity. Materials Advances. 2020;**1**(5):1193-1201

[66] Sudrajat H, Babel S, Ta AT, Nguyen TK. Mn-doped TiO_2 photocatalysts: Role, chemical identity, and local structure of dopant. Journal of Physics and Chemistry of Solids. 2020;**144**:109517

[67] Xochihua Juan JL, Solis Maldonado C, Luna Sánchez RA, Enciso Díaz OJ, Rojas Ronquillo MR, Sandoval-Rangel L, et al. TiO_2 doped with europium (Eu): Synthesis, characterization and catalytic performance on pesticide degradation under solar irradiation. Catalysis Today. 2022;**394-396**:304-313

[68] Orizu GE, Ugwuoke PE, Asogwa PU, Offiah SU. A review on the inference of doping TiO_2 with metals/non-metals to improve its photocatalytic activities. IOP Conference Series Earth and Environmental Science. 2023;**1178**(1):012008

[69] Chen X, Liu L, Yu PY, Mao SS. Increasing solar absorption for photocatalysis with black hydrogenated titanium dioxide nanocrystals. Science. 2011;**331**(6018):746-750

[70] Naldoni A, Altomare M, Zoppellaro G, Liu N, Kment Š, Zbořil R, et al. Photocatalysis with reduced TiO_2: From black TiO_2 to cocatalyst-free hydrogen production. ACS Catalysis. 2019;**9**(1):345-364

[71] Asiah MN, Mamat MH, Khusaimi Z, Abdullah S, Rusop M, Qurashi A. Surfactant-free seed-mediated large-scale synthesis of mesoporous TiO_2 nanowires. Ceramics International. 2015;**41**(3, Part A):4260-4266

[72] Ye M, Chen Z, Wang W, Shen J, Ma J. Hydrothermal synthesis of TiO_2 hollow

microspheres for the photocatalytic degradation of 4-chloronitrobenzene. Journal of Hazardous Materials. 2010;**184**(1):612-619

[73] Liu Z, Sun DD, Guo P, Leckie JO. One-step fabrication and high photocatalytic activity of porous TiO_2 hollow aggregates by using a low-temperature hydrothermal method without templates. Chemistry—A European Journal. 2007;**13**(6):1851-1855

[74] Pan JH, Wang XZ, Huang Q, Shen C, Koh ZY, Wang Q, et al. Large-scale synthesis of urchin-like mesoporous TiO_2 hollow spheres by targeted etching and their photoelectrochemical properties. Advanced Functional Materials. 2014;**24**(1):95-104

[75] Li X, Xiong Y, Li Z, Xie Y. Large-scale fabrication of TiO_2 hierarchical hollow spheres. Inorganic Chemistry. 2006;**45**(9):3493-3495

[76] Sha D, Wang J, Wu X, Zou H, Dai Y, Ren J, et al. One-step and large-scale preparation of TiO_2/amorphous carbon composites with excellent visible light photocatalytic properties. RSC Advances. 2016;**6**(70):65607-65612

[77] Wu H, Qin P, Cao S, Luo G, Wang C, Tu R, et al. Large-scale synthesis of size-controlled amorphous and anatase TiO_2 via a benzoic acid-assisted sol-gel-hydrothermal process. Ceramics International. 2024;**50**(12):21889-21897

[78] Petronella F, Truppi A, Dell'Edera M, Agostiano A, Curri ML, Comparelli R. Scalable synthesis of mesoporous TiO_2 for environmental photocatalytic applications. Materials. 2019;**12**(11):1853

[79] A short survey | www.nanothecaba.unito.it [Internet]. [cited 2024 May 31]. Available from: http://www.nanothecaba.unito.it/

[80] Marques JA, Costa PG, Marangoni LFB, Pereira CM, Abrantes DP, Calderon EN, et al. Environmental health in southwestern Atlantic coral reefs: Geochemical, water quality and ecological indicators. Science of the Total Environment. 2019;**651**:261-270

[81] Ângelo J, Andrade L, Madeira LM, Mendes A. An overview of photocatalysis phenomena applied to NOx abatement. Journal of Environmental Management. 2013;**129**:522-539

[82] Chong MN, Jin B, Chow CWK, Saint C. Recent developments in photocatalytic water treatment technology: A review. Water Research. 2010;**44**(10):2997-3027

[83] Ullah S, Ferreira-Neto EP, Pasa AA, Alcântara CCJ, Acuña JJS, Bilmes SA, et al. Enhanced photocatalytic properties of core@shell SiO_2@TiO_2 nanoparticles. Applied Catalysis B: Environmental. 2015;**179**:333-343

[84] Lee JW, Kong S, Kim WS, Kim J. Preparation and characterization of SiO_2/TiO_2 core-shell particles with controlled shell thickness. Materials Chemistry and Physics. 2007;**106**(1):39-44

[85] Huang C, Bai H, Huang Y, Liu S, Yen S, Tseng Y. Synthesis of neutral SiO_2/TiO_2 hydrosol and its application as antireflective self-cleaning thin film. International Journal of Photoenergy. 2012;**2012**(1):620764

[86] Babyszko A, Wanag A, Sadłowski M, Kusiak-Nejman E, Morawski AW. Synthesis and characterization of SiO_2/TiO_2 as photocatalyst on methylene blue degradation. Catalysts. 2022;**12**(11):1372

[87] Ekka B, Sahu MK, Patel RK, Dash P. Titania coated silica nanocomposite prepared via encapsulation method for

the degradation of safranin-O dye from aqueous solution: Optimization using statistical design. Water Resources and Industry. 2019;**22**:100071

[88] Budiarti HA, Puspitasari RN, Hatta AM, Sekartedjo RDD. Synthesis and characterization of TiO_2@SiO_2 and SiO_2@TiO_2 core-shell structure using Lapindo mud extract via sol-gel method. Procedia Engineering. 2017;**170**:65-71

[89] Wang W, Meng Z, Zhang Q, Jia X, Xi K. Synthesis of stable Au–SiO_2 composite nanospheres with good catalytic activity and SERS effect. Journal of Colloid and Interface Science. 2014;**418**:1-7

[90] Utami FD, Rahman DY, Margareta DO, Rahmayanti HD, Munir R, Sustini E, et al. TiO_2 photocatalytic degradation of methylene blue using simple spray method. IOP Conference Series Material Science Engineering. 2019;**599**(1):012026

[91] de Souza PM, Mendes RP, Bellettini GC, Benetti RM, Elyseu F, Bernardin AM. Photocatalytic discoloration of methylene blue by TiO_2 P25 under UV light using ISO 10678 standard as a guide. Journal of Photochemistry and Photobiology Chemistry. 2023;**435**:114304